VIBRATION OF FUNCTIONALLY GRADED BEAMS AND PLATES

VIBRATION OF FUNCTIONALLY GRADED BEAMS AND PLATES

SNEHASHISH CHAKRAVERTY

KARAN KUMAR PRADHAN

Department of Mathematics, National Institute of Technology Rourkela, Odisha, India

ELSEVIER

AMSTERDAM • BOSTON • HEIDELBERG • LONDON
NEW YORK • OXFORD • PARIS • SAN DIEGO
SAN FRANCISCO • SINGAPORE • SYDNEY • TOKYO
Academic Press is an imprint of Elsevier

Academic Press is an imprint of Elsevier
125 London Wall, London, EC2Y 5AS, UK
525 B Street, Suite 1800, San Diego, CA 92101-4495, USA
50 Hampshire Street, 5th Floor, Cambridge, MA 02139, USA
The Boulevard, Langford Lane, Kidlington, Oxford OX5 1GB, UK

Notices
Knowledge and best practice in this field are constantly changing. As new research and experience broaden our understanding, changes in research methods, professional practices, or medical treatment may become necessary.

Practitioners and researchers must always rely on their own experience and knowledge in evaluating and using any information, methods, compounds, or experiments described herein. In using such information or methods they should be mindful of their own safety and the safety of others, including parties for whom they have a professional responsibility.

To the fullest extent of the law, neither the Publisher nor the authors, contributors, or editors, assume any liability for any injury and/or damage to persons or property as a matter of products liability, negligence or otherwise, or from any use or operation of any methods, products, instructions, or ideas contained in the material herein.

British Library Cataloguing in Publication Data
A catalogue record for this book is available from the British Library

Library of Congress Cataloging-in-Publication Data
A catalog record for this book is available from the Library of Congress

For information on all Academic Press publications visit our website at http://store.elsevier.com/

ISBN: 978-0-12-804228-1

CONTENTS

CHAPTER 1

Introduction

This chapter includes the description on structural beams and plates and then a brief discussion on composites. Next, the concept of functionally graded (FG) composites and vibration preliminaries along with their classifications are incorporated. Finally, the terminologies associated with vibration problems, viz. literature positions related to beam (and plate) theories and boundary conditions, are discussed.

1.1 STRUCTURAL BEAMS AND PLATES

In common practice, dynamic systems may be categorized into two classes: discrete and continuous (or distributed). Discrete systems are defined by the variables that depend only on time, whereas continuous systems are described by variables that depend on time and space. In this regard, the equations of motion of discrete systems are governed by ordinary differential equations (ODEs) and those of continuous systems by partial differential equations (PDEs). The mass and elasticity in structural components, viz. strings, beams, rods, cables, plates, and shells of mechanical or structural systems, are assumed to be distributed in specified domains and hence these are referred to as continuous (or distributed) systems. In other words, corresponding structural components may be considered as continuum systems which have infinite number of Degrees of Freedom (DOF[1]), which involve variables that depend on time as well as the spatial coordinates.

1.2 BACKGROUND OF COMPOSITES

In general, the engineering materials can be classified into three primary classes: metals and alloys; ceramics and glasses; and polymers. Among these three primary classes, metals, metallic alloys, and polymers are more widely

[1]Degrees of freedom is defined as the minimum number of independent variables to describe fully the motion of a system.

Vibration of Functionally Graded Beams and Plates
http://dx.doi.org/10.1016/B978-0-12-804228-1.00001-8

implemented in various structural and engineering applications than ceramics and glasses. Somehow, ceramics have also drawn considerable attention from the scientific community in last three decades. The metals and metallic alloys are best known for their high tensile strength and high toughness (crack growth resistance), as well as their ability to be manufactured in various geometries. Ceramics and glasses have advantageous properties, including refractoriness (capability to withstand high temperatures), strength retention at high temperature, a high melting point, and good mechanical properties (hardness, elastic modulus, and compressive strength). Polymers have distinct advantages because of their low density, high flexibility, and ability to be molded into different shapes and sizes. In view of such an attractive combination of properties, ceramics are considered as potential materials for high-temperature structural applications and various tribological applications requiring high hardness and wear resistance. It is rightly mentioned in Basu and Balani (2011) that a derived material class (composites) can be developed by combining various advantageous properties of three primary material classes. The composites are generally defined as a class of materials that comprise of two or more intimately bonded microstructural phases, aimed to provide better properties than those of individual components. Depending on whether metals, ceramics, or polymers comprise more than 50% by volume of a composite, it can further be classified as a Metal Matrix Composite (MMC), a Ceramic Matrix Composite (CMC), or a Polymer Matrix Composite (PMC), respectively. Moreover, a CMC so-called functionally graded composite (FGC) has been of considerable importance in recent years because of its thermal resistance in environments where temperatures fluctuate to extremes.

1.3 VIBRATION PRELIMINARIES

As mentioned in Kelly (1999), vibrations are oscillations (or fluctuations) of the mechanical and structural systems around an equilibrium position. The motion of a guitar string, motion felt by passengers in an automobile traveling over a bumpy road, or swaying of tall buildings due to an earthquake or motion of an airplane in turbulence are typical examples of vibration. When an inertia element is deflected from its equilibrium position due to an external energy applied to the system, vibrations are initiated. With certain assumptions, the oscillations in vibration problems may be described by simple harmonic motion. This simple harmonic motion may be described by periodic oscillation of the object (or continuous system) around the equilibrium position. The maximum displacement from equilibrium position is referred to as the amplitude and is assumed to be constant in simple harmonic motion. Vibration in mechanical or structural systems may be classified into the following types (Rao, 2007):

- Undamped vibration: The type of vibration is undamped if there is no loss or dissipation of energy due to friction or other resistance during vibration of a system.
- Damped vibration: The system is considered damped if energy dissipation occurs in the presence of damping components during vibration. Although finding dynamic characteristics is easier and simpler if neglecting damping, a consideration of damping becomes extremely important if the system operates near resonance.
- Free vibration: If a system oscillates only due to an initial disturbance, the system is said to undergo free vibration. Here, no external force is applied after the initial disturbance (or after time zero).
- Forced vibration: The system is said to be under forced vibration if the system vibrates due to the application of an external force.
- Linear vibration: If all the basic components (mass, spring, and damper) of a vibrating system act linearly, the resulting vibration is called linear vibration. The governing equation of motion in linear vibration must be a linear differential equation.
- Nonlinear vibration: If any of the basic components of a vibrating system behave nonlinearly, the nonlinear vibration occurs. The equation governing nonlinear vibration will be a nonlinear differential equation.
- Self-excited vibration: This is a periodic and deterministic oscillation. There are systems in which the exciting force is a function of the motion variables (displacement, velocity, or acceleration) and thus it varies with the motion it produces. A few illustrations of self-excited vibration are friction-induced vibration (in vehicle clutches and brakes, vehicle-bridge interaction), flow-induced vibration (circular wood saws, CDs, DVDs, in machining, fluid-conveying pipelines), and undesired oscillation of an airplane wing.

The study on vibration of beams and plates is an extremely important area owing to the wide variety of applications involved. Since these structural members form integral parts of structures, prior knowledge of their vibration behavior is essential for an engineer before finalizing the design of a given structure. In particular, beams and plates with different shapes subjected to boundary conditions at the edges are often encountered in several engineering applications such as aeronautical engineering, automobile and telephone industries, machine design, nuclear reactor technology, naval structures, and earthquake-resistant structures. Accordingly, the vibration characteristics may be evaluated with the help of analytical and different computational techniques.

In general, the laws of the universe are written in the language of mathematics. Algebra is sufficient to solve many static problems, but the most important natural phenomena involve change and are described by equations that relate changing quantities. These are generally referred to as differential equations (Penny, 2004). In a similar fashion, vibration problems associated with structural members are governed by higher-order PDEs. However, it is not always possible to find the analytical solutions for such problems. Accordingly, numerical methods may be applied to solve such differential equations. Although various numerical methods exist to handle these PDEs, these are sometimes problem dependent and may not handle all sets of boundary conditions, along with complicating effects, with ease. A brief idea on theoretical formulation associated with mechanics of different structural members may be found in the available resources (Timoshenko and Woinowsky-Krieger, 1959; Wang et al., 2000; Reddy, 2000; Rao, 2004; Bhavikatti, 2005; Chakraverty, 2009) and the literatures provided therein. In earlier decades, the isotropic structural members along with composites have played crucial role in different engineering applications, structural design, and architectures. On the contrary, the thermal withstanding behavior of FG structural members have become significant in these applications. It can easily be noted in Abrate (2008) that FG plates behave like homogeneous plates that require no special tools to analyze their mechanics. It is therefore a challenging task to develop numerical methods that may be general to undertake the investigation. As such, the present study aims to develop mathematical models based on different computational procedures for vibration of FG structural beams and plates to obtain the corresponding generalized eigenvalue problems.

1.3.1 Beam and Plate Theories

Generally in vibration problems, the displacement fields of deformed beam (or plate) may be decided by shear deformation beam (or plate) theories. Instead of classical beam (or plate) theory, different forms of deformation theories may also be observed in open literature. It may occur due to the fact that classical beam (or plate) theory neglects transverse shear deformation effects. Based on certain assumptions given in Reddy (1984) and Aydogdu (2009), one may easily provide alternate forms of shear deformation theories.

Accordingly, a list of deformation beam and plate theories related to isotropic members is available in Wang et al. (2000). Further, Reddy (1984) has given a refined nonlinear theory accounting for the von Karman strains to find exact solutions for simply supported plates. A new higher-order shear deformable laminated composite plate theory is proposed by Aydogdu (2009) using 3D elasticity bending solutions in an inverse method. Infinitesimal deformations of a homogeneous thick elastic plate have been investigated by Xiao et al. (2007), using a meshless Petrov-Galerkin (MLPG) method and a higher-order shear and normal deformable plate theory. Moreover, Reddy (2011) has developed microstructure-dependent nonlinear Euler-Bernoulli and Timoshenko beam theories using the principle of virtual displacements. A general assessment of inverse trigonometric shear deformation theory is developed by Grover et al. (2013) to analyze structural responses of laminated-composite and sandwich plates. A general higher-order shear deformation theory is proposed by Qu et al. (2013) for free and transient vibration analyses of composite laminated beams with arbitrary combinations of classical and non-classical boundary conditions. Thai et al. (2014) have recently introduced an inverse tangent shear deformation theory for the static, free vibration and buckling analysis of laminated composite and sandwich plates. In a similar fashion, a major topic in deriving shear deformation theories can also be found in Sina et al. (2009), Xiang et al. (2009), Şimşek (2010a), Thai and Vo (2012), Şimşek and Reddy (2013), Qu et al. (2013), Vo et al. (2013) and the literatures mentioned therein.

1.3.2 Boundary Conditions

Owing to the solution of governing PDEs of continuous systems, we may need the value of the dependent variable and probably their derivatives at more than one point or on the boundary. These problems are said to be boundary value problems (BVPs). In particular, this book considers three classical boundary conditions, viz. clamped, simply supported, and free, and the corresponding conditions are mentioned below:

- Clamped: This case considers both displacement and slope (or rotation) to be zero.
- Simply supported: The displacement and bending moment must be zero in this boundary condition.
- Free: Here, the bending moment and shear force must be zero.

In addition, the solution also depends on certain assumptions and on the appropriate choice of a shape or trial function, which is dependent on the physical characteristics of the system. The selected shape functions should satisfy certain boundary conditions of the problem. Thus we may find two classes of boundary condition: essential (or geometric) and natural (or dynamic) boundary conditions. These conditions may be defined as follows:

- Essential: These boundary conditions are demanded by the displacements or slopes on the boundary of a physical body. Essential boundary conditions are also referred to as Dirichlet boundary conditions.
- Natural: These conditions are demanded by the condition of bending moment and shearing force balance. As such, natural boundary conditions are also known as Neumann boundary conditions.

Based on essential and natural boundary conditions, the BVPs may also be divided into three categories: Dirichlet, Neumann, and mixed. Problems in which all the boundary conditions are of the essential type are called Dirichlet problems, and those in which the boundary conditions are of the natural type are said to be Neumann problems. On the other hand, the problem known as mixed type assumes both essential and natural boundary conditions at the edges.

CHAPTER 2

Origin and Basics of Functionally Graded Structural Members

In this chapter, a short history of functionally graded (FG) composites along with their applications in several fields are discussed. Then the concept and variation of material properties in the case of the FG beam and plate members are given, followed by previously proposed shear deformation beam (and plate) theories. Rather than assuming only existing theories, a few new deformation beam theories are also introduced based on certain basic assumptions. Moreover, the governing equations associated with the titled problems are also reported by considering the deformation theories.

2.1 HISTORY AND APPLICATION OF FG COMPOSITES

The concept of functionally graded materials (FGMs) was first enunciated in 1984 by a group of material scientists in Japan during a space plane project in the form of thermal barrier material which can withstand a huge temperature fluctuation across a very thin cross-section (Koizumi, 1993; Loy et al., 1999). Since then, FGMs have received major attention as heat-shielding advanced structural materials in various engineering applications and in manufacturing industries, viz. aerospace, nuclear reactor, automobile, aircrafts, space vehicles, and biomedical and steel industries. FGMs were initially designed as thermal barrier materials for aerospace structures and fusion reactors (Hirai and Chen, 1999). The current focus of materials development activities for both composites and FGMs includes improvements in material performance, ability to support optimized structural designs, continued lowering of manufacturing costs, and ability to perform reliably in service (Udupa et al., 2012). Two different possibilities of gradation behavior may be observed in the case of FG composites (Shen, 2009): (i) continuous gradation of two different material phases from one interface to another and (ii) the change of material phases in a discontinuous (or stepwise) pattern is called segmented FGMs.

Vibration of Functionally Graded Beams and Plates
http://dx.doi.org/10.1016/B978-0-12-804228-1.00002-X

Due to gradation of material properties and thermal resistance behavior, there are a wide variety of applications of these materials in different sectors (Miyamoto et al., 1999):

- Automobiles: Combustion chambers (SiC-SiC), engine cylinder liners (Al-SiC), CNG storage cylinders, diesel engine pistons (SiCw/Al-alloy), brake rotors, leaf springs (E-glass/epoxy), drive shafts (Al-C), motorcycle drive sprocket, pulleys, torque converter reactors, shock absorbers (SiCp/Al-alloy), radiator end caps.
- Sub-marine: Propulsion shaft (carbon and glass fibers), cylindrical pressure hull (graphite/epoxy), sonar domes (glass/epoxy), composite piping systems, scuba diving cylinders (Al-SiC), floats, boat hulls.
- Commercial and industrial: Pressure vessels, fuel tanks, cutting tool inserts, laptop cases, wind turbine blades, electric motors, firefighting air bottles, artificial ligaments, MRI scanner cryogenic tubes, wheelchairs, hip joint implants, eyeglass frames, camera tripods, musical instruments, drilling motor shafts, drill casings, crane components, high-pressure hydraulic pipes, X-ray tables, heart valves, helmets, crucibles, beams.
- Aerospace equipment and structures: Rocket nozzles (TiAl-SiC fibers), heat exchanger panels, engine parts (Be-Al), wind tunnel blades, spacecraft truss structures, reflectors, solar panels, camera housings, hubble space telescope metering truss assembly, turbine wheels (operating above 40,000 rpm), nose caps, and leading edge of missiles and space shuttle.
- Aerospace: Wings, rotary launchers, engine casings, rings (Al_2O_3/Al-alloy), drive shafts, propeller blades, landing gear doors, thrust reversers (carbon/bismaleimide), helicopter components, viz. rotor drive shafts, mast mounts, main rotor blades (carbon/epoxy).
- Sports: Racing bicycle frames (SiCw/6061), racing vehicle frames.

2.2 VARIATION OF MATERIAL PROPERTIES

The FG materials are generally ceramic-metal composites in which material properties vary continuously in thickness direction from one interface to another in a specific mathematical pattern, thus eliminating interface problems and mitigating thermal stress concentrations. The ceramic constituent provides high-temperature resistance due to its low thermal conductivity, whereas the ductile metal constituent prevents fracture caused by stresses due to high temperature gradient in a very short span of time and provides stronger mechanical performance. The continuous variation of material properties of FG beam (or plate) constituents in power law can be defined as (Aydogdu and Taskin, 2007; Şimşek and Kocatürk, 2009; Sina et al., 2009)

$$\mathcal{P}(z) = (\mathcal{P}_c - \mathcal{P}_m)\left(\frac{z}{h} + \frac{1}{2}\right)^k + \mathcal{P}_m \tag{2.1}$$

and on the other hand, in exponential law, as (Aydogdu and Taskin, 2007; Şimşek and Kocatürk, 2009)

$$\mathcal{P}(z) = \mathcal{P}_c e^{-\delta\left(1 - \frac{2z}{h}\right)} \tag{2.2}$$

In Eqs. (2.1) and (2.2), $\mathcal{P}_c$ and $\mathcal{P}_m$ denote the values of the material properties of the ceramic and metal constituents of the FG beam (or plate), respectively. In power-law variation, k (power-law exponent) is a non-negative variable parameter. In exponential gradation, $\delta = \frac{1}{2}\ln\left(\frac{\mathcal{P}_c}{\mathcal{P}_m}\right)$. According to these gradation behaviors, the bottom surface ($z = -h/2$) of FG beam (or plate) is pure metal, whereas the top surface ($z = h/2$) is pure ceramic, and for different values of k one can obtain different volume fractions of material beam (or plate) as mentioned in Aydogdu and Taskin (2007). For our present formulations, the material properties, viz. Young's modulus (E) and mass density (ρ), are taken to vary along thickness direction except Poisson's ratio (ν) remains constant (Fig. 2.1).

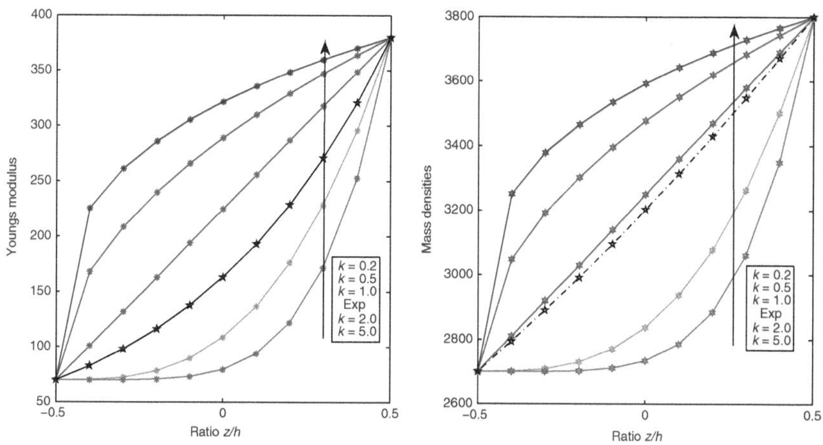

(a) Variation of Young's modulus (b) Variation of mass densities

Figure 2.1 Power-law and exponential variation of (a) Young's modulus and (b) mass densities of the FG beam (or plate). (*Reproduced from Chakraverty, S., Pradhan, K.K., Free vibration of exponential functionally graded rectangular plates in thermal environment with general boundary conditions. Aerospace Science and Technology 2014;36:132–156. Copyright © 2015 Elsevier Masson SAS. All rights reserved.*).

2.3 BEAM AND PLATE THEORIES

To study the vibration characteristics, the displacement field(s) of deformed FG beams and plates are considered with respect to transverse displacement. Higher-order deformation beam theories of different forms are assumed by taking transverse shear deformations in the case of FG beams, whereas the classical plate theory (CPT) is considered in the case of FG plates with different geometries. In this regard, the beam and plate theories along with their constitutive relations are outlined as below.

2.3.1 Higher-Order Shear Deformation Beam Theory

A straight FG beam of length L, width b, and thickness h, having rectangular cross-section with cartesian coordinate system $O(x, y, z)$ having the origin at O as shown in Fig. 2.2.

The dynamic problems concerned with FG beam are to be handled by the shear deformation beam theories (SDBTs). As such, let the deformation of FG beam be assumed in the x–z plane and denote the displacement components along x, y, and z directions by u_x, u_y, and u_z, respectively. To derive the numerical model, the following assumptions are made for the displacement field (Vo et al., 2013):

- The axial and transverse displacements consist of bending and shear components, in which the bending components do not contribute toward shear forces and likewise, the shear components do not contribute toward bending moments.
- The bending component of axial displacement is similar to that given by the Euler-Bernoulli beam theory.

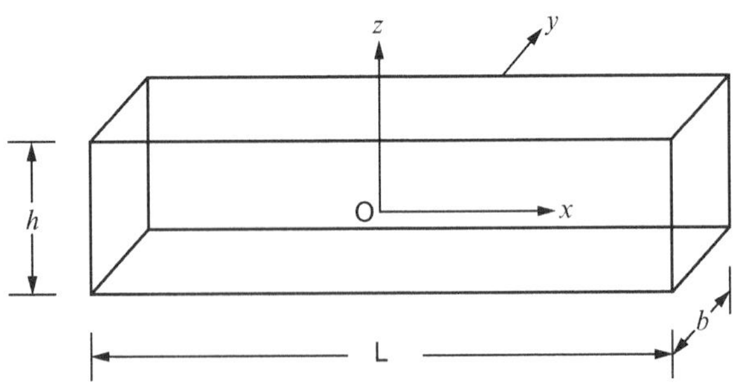

Figure 2.2 A typical functionally graded beam element with cartesian coordinates. *(Reprinted from International Journal of Mechanical Sciences, Vol 82, Pradhan, K.K., Chakraverty, S., Effects of different shear deformation theories on free vibration of functionally graded beams, Pages 149–160, Copyright (2014), with permission from Elsevier.).*

- The shear component of axial displacement gives rise to the higher-order variation of shear strain and hence to shear stress through the beam depth in such a way that shear stress vanishes on the top and bottom surfaces.

Based on the higher-order SDBT, the axial displacement (u_x) and the transverse displacement (u_z) at any point of the beam are given as below (Aydogdu and Taskin, 2007):

$$\begin{aligned} u_x(x,z) &= u(x,t) - zw_{,x}(x,t) + f(z)v(x,t) \\ u_z(x,z) &= w(x,t) \end{aligned} \tag{2.3}$$

where u and w represent the axial and the transverse displacement of any point on the neutral axis, respectively, while v is an unknown function that represents the effect of transverse shear strain on the neutral axis. $f(z)$ represents the shape function determining the distribution of the transverse shear stress and strain through the thickness of the beam and $(\ )_{,x}$ indicates the derivative with respect to x. Different theories can be obtained by choosing their respective shape functions $f(z)$. The present study is concerned with SDBTs, viz. classical beam theory (CBT), Timoshenko beam theory (TBT), parabolic shear deformation beam theory (PSDBT), exponential shear deformation beam theory (ESDBT), trigonometric shear deformation beam theory (TSDBT), hyperbolic shear deformation beam theory (HSDBT), and a new shear deformation beam theory (ASDBT), as mentioned in Şimşek (2010a). $f(z)$ for these SDBTs are given below (Şimşek, 2010a):

$$\begin{aligned} &\text{CBT}: f(z) = 0 \\ &\text{TBT}: f(z) = z \\ &\text{PSDBT}: f(z) = z\left(1 - \frac{4z^2}{3h^2}\right) \\ &\text{ESDBT}: f(z) = z e^{-2(z/h)^2} \\ &\text{HSDBT}: f(z) = h\sinh\left(\frac{z}{h}\right) - z\cosh\left(\frac{1}{2}\right) \\ &\text{TSDBT}: f(z) = \frac{h}{\pi}\sin\left(\frac{\pi z}{h}\right) \\ &\text{ASDBT}: f(z) = z\alpha^{-2(z/h)^2/\ln\alpha} \quad \text{with } \alpha = 3 \end{aligned} \tag{2.4}$$

After observing Eqs. (2.3) and (2.4), it is evident that the displacement field for the FG beam based on a general higher-order shear deformation theory may be decided by the function $f(z)$ and the major criteria to choose the shape functions ($f(z)$) for these models are given as follows (Reddy, 1984; Aydogdu, 2009):

- Suitable $f(z)$ function should approximately satisfy parabolic shear deformation distribution.
- The boundary conditions should be satisfied on the bottom and top surfaces of the beam.

Accordingly, the authors have proposed two different types of shear deformation theories which involve inverse trigonometric functions and a generalized power-law exponent based function. These theories may be named as inverse sine shear deformation beam theory (ISDBT), inverse cosine shear deformation beam theory (ICDBT), inverse tangent shear deformation beam theory (ITDBT), and power-law exponent based shear deformation beam theory (PESDBT).

$$\text{ISDBT} : f(z) = \frac{2z}{\sqrt{3}} - h\arcsin\left(\frac{z}{h}\right)$$

$$\text{ICDBT} : f(z) = \frac{2z}{\sqrt{3}} + h\arccos\left(\frac{z}{h}\right)$$

$$\text{ITDBT} : f(z) = \frac{4z}{5} - h\arctan\left(\frac{z}{h}\right)$$

$$\text{PESDBT} : f(z) = h\left(\frac{z}{h}\right)^{2n+1} - (2n+1)z\left(\frac{1}{2}\right)^{2n} \tag{2.5}$$

The derivatives of $f(z)$ in the proposed inverse trigonometric SDBTs yield

$$\frac{\partial f}{\partial z} = \begin{cases} \frac{2}{\sqrt{3}} - \frac{1}{\sqrt{1-\left(\frac{z}{h}\right)^2}}; & \text{for ISDBT and ICDBT} \\ \frac{4}{5} - \frac{1}{1+\left(\frac{z}{h}\right)^2}; & \text{for ITDBT} \end{cases}$$

The transverse shear stress on the bottom and top surfaces of the beam may be found as

$$\tau_{xz}\left(x, \mp\frac{h}{2}\right) = 0 \tag{2.6}$$

On the contrary, n is a non-negative integer which will play a key role in PESDBT. It can be viewed that classical or Euler-Bernoulli beam theory is a special case of this model. If we assume $n = 0$, then PESDBT will follow the assumptions of CBT. We can define the limiting value of $f(z)$ associated with PESDBT for very large value of n as

$$\begin{aligned}
\lim_{n\to+\infty} f(z) &= \lim_{n\to+\infty}\left\{h\left(\frac{z}{h}\right)^{2n+1} - (2n+1)z\left(\frac{1}{2}\right)^{2n}\right\}\\
&= \lim_{n\to+\infty} h\left(\frac{z}{h}\right)^{2n+1} - \lim_{n\to+\infty}(2n+1)z\left(\frac{1}{2}\right)^{2n}\\
&= z\left\{\lim_{n\to+\infty}\left(\frac{z}{h}\right)^{2n} - \lim_{n\to+\infty}\left(\frac{2n+1}{2^{2n}}\right)\right\},\\
&\text{where } -\frac{h}{2}\le z\le+\frac{h}{2}\\
\Rightarrow \lim_{n\to+\infty} f(z) &= 0 \quad \text{(by Squeeze theorem)}
\end{aligned} \tag{2.7}$$

As expressed in Eq. (2.7), it can easily be noticed that limiting value of $f(z)$ approaches to zero for very large value of n. So classical or Euler-Bernoulli beam theory can also be derived from the proposed shear deformation theory by assuming higher values of n. It is also easy to satisfy the stress free boundary conditions at the top and bottom surfaces of the beam in PESDBT by simply finding the derivative in this case. The constitutive relations in Eqs. (2.10) and (2.11) given for existing beam theories will also be similar to the proposed beam theories except the change in shape functions.

The following relations represent the kinematic relations as per the above displacement field assumed in Eq. (2.3)

$$\varepsilon_{xx} = u_{,x} - zw_{,xx} + f(z)v_{,x} \tag{2.8}$$

$$\gamma_{xz} = f'(z)v \tag{2.9}$$

where a prime denotes the derivative with respect to z. ε_{xx} and γ_{xz} are the normal and shear strains, respectively. By assuming the material constituents of FGM beam to obey the generalized Hooke's law, the state of stresses in the beam can be expressed as

$$\sigma_{xx} = Q_{11}\varepsilon_{xx} \tag{2.10}$$

$$\tau_{xz} = Q_{55}\gamma_{xz} \tag{2.11}$$

where σ_{xx} and τ_{xz} are the normal and the shear stresses, respectively, and Q_{ij} are the transformed stiffness constants in the beam coordinate system, which are defined as

$$Q_{11} = \frac{E(z)}{1-\nu^2}, \quad Q_{55} = \frac{E(z)}{2(1+\nu)}$$

2.3.2 Classical Plate Theory

This book considers the CPT in the analysis along with the solution of FG plates of various shapes, viz. rectangular, elliptic, and triangular along with different complicating effects (elastic foundation, thermal environment, and piezoelectricity). Here, we shall address the CPT and its assumptions will be outlined for the sake of completeness. Classical or Kirchoff's plate theory is based on the following assumptions (Timoshenko and Woinowsky-Krieger, 1959; Wang et al., 2000; Chakraverty, 2009):

- The thickness of the plate is small when compared with other dimensions.
- The normal stresses in the direction transverse to the plate are taken to be negligibly small.
- The effect of rotatory inertia is negligible. Normal to the undeformed middle surface remains straight and normal to the deformed middle surface and unstretched in length.

Figure 2.3 represents three different types of geometries, viz. rectangular, elliptic, and triangular assumed in vibration problems related to the FG plates. Irrespective of geometry, CPT for the displacement field of FG plate can be expressed as (Wang et al., 2000)

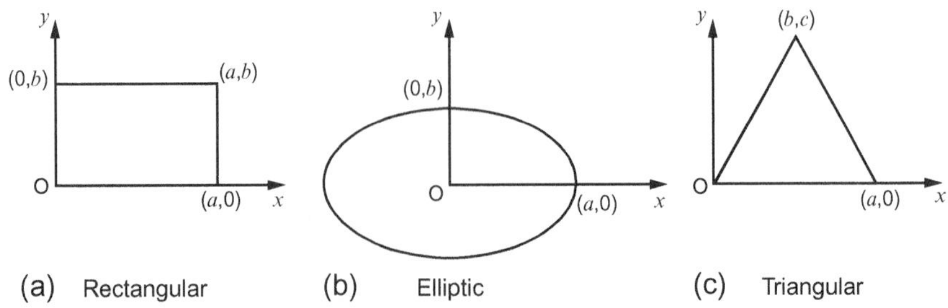

Figure 2.3 Typical functionally graded rectangular, elliptic, and triangular plate elements: (a) Rectangular; (b) Elliptic; (c) Triangular.

$$
\begin{aligned}
u_x(x, y, z) &= -z\frac{\partial w}{\partial x} \\
u_y(x, y, z) &= -z\frac{\partial w}{\partial y} \\
u_z(x, y, z) &= w(x, y)
\end{aligned}
\tag{2.12}
$$

where u_x, u_y, and u_z are the displacement components along x, y, and z coordinate directions, respectively, and w is the transverse deflection at a point on the midplane (x–y plane). The transverse shear deformation has been neglected in the case of Kirchoff's assumption that is deformation is due to bending and in-plane stretching only. The non-zero linear strains associated with the displacement field can be expressed as

$$
\begin{Bmatrix} \epsilon_{xx} \\ \epsilon_{yy} \\ \gamma_{xy} \end{Bmatrix} = \begin{Bmatrix} \frac{\partial u_x}{\partial x} \\ \frac{\partial u_y}{\partial y} \\ \frac{\partial u_x}{\partial y} + \frac{\partial u_y}{\partial x} \end{Bmatrix} = \begin{Bmatrix} -z\frac{\partial^2 w}{\partial x^2} \\ -z\frac{\partial^2 w}{\partial y^2} \\ -2z\frac{\partial^2 w}{\partial x \partial y} \end{Bmatrix}
\tag{2.13}
$$

where ϵ_{xx} and ϵ_{yy} are the normal strains in x and y directions, respectively, and γ_{xy} is the shear strain. By assuming the material constituents of FG plate to obey the generalized Hooke's law, the stress-strain relationships can be expressed in matrix form as

$$
\begin{Bmatrix} \sigma_{xx} \\ \sigma_{yy} \\ \tau_{xy} \end{Bmatrix} = \begin{pmatrix} Q_{11} & Q_{12} & 0 \\ Q_{21} & Q_{22} & 0 \\ 0 & 0 & Q_{66} \end{pmatrix} \begin{Bmatrix} \epsilon_{xx} \\ \epsilon_{yy} \\ \gamma_{xy} \end{Bmatrix}
\tag{2.14}
$$

where σ_{xx}, σ_{yy} are the normal stresses, τ_{xy} is the shear stress and the reduced stiffness components, and Q_{ij} $(i, j = 1,\ 2,\ 6)$ are given by

$$
Q_{11} = Q_{22} = \frac{E(z)}{1-\nu^2}, \quad Q_{12} = Q_{21} = \frac{\nu E(z)}{1-\nu^2}, \quad Q_{66} = \frac{E(z)}{2(1+\nu)}
$$

Here E and ν are Young's modulus and Poisson's ratio of the material constituents, respectively.

2.4 GOVERNING EQUATIONS

Based on the displacement fields of deformed FG beams and plates, the governing equations related to the vibration problems may be found by using Hamilton's principle. Detailed procedure of finding Lagrangian function and fundamental lemma of calculus of variations to develop the Euler-Lagrange equations have not been discussed here, but can be found in most of the open resources (Sina et al., 2009; Nie et al., 2013; Mahi et al., 2010) and the literatures provided therein.

2.4.1 Vibration of FG Beam

The governing differential equations defining free vibration of FG beam within the framework of general higher-order shear deformation theory can be written as (Aydogdu and Taskin, 2007)

$$
\begin{aligned}
\frac{\partial N_x}{\partial x} &= \rho_0 \ddot{u} + \rho_{01}\ddot{v} - \rho_1 \frac{\partial \ddot{w}}{\partial x} \\
\frac{\partial^2 M_x}{\partial x^2} &= \rho_1 \frac{\partial \ddot{u}}{\partial x} + \rho_{11}\frac{\partial \ddot{v}}{\partial x} + \rho_0 \ddot{w} - \rho_2 \frac{\partial^2 \ddot{w}}{\partial x^2} \\
\frac{\partial M_x^f}{\partial x} - Q_x^f &= \rho_{01}\ddot{u} + \rho_{02}\ddot{v} - \rho_{11}\frac{\partial \ddot{w}}{\partial x}
\end{aligned} \tag{2.15}
$$

where $\ddot{( )}$ denotes the second time derivatives. Here ρ_i and ρ_{jm} are the inertial coefficients, where $\rho_i = \int_{-h/2}^{h/2} \rho(z) z^i \, \mathrm{d}z$, $i = 0, 1, 2$; $\rho_{jm} = \int_{-h/2}^{h/2} \rho(z) z^j f^m \, \mathrm{d}z$, $j = 0, 1$; $m = 1, 2$ and the force and moment components can be expressed as

$$
\begin{aligned}
\left(N_x, M_x, M_x^f\right) &= \int_{-h/2}^{h/2} \sigma_{xx}\left(1,\ z,\ f(z)\right) \mathrm{d}z \\
Q_x^f &= \int_{-h/2}^{h/2} \tau_{xz} f'(z)\, \mathrm{d}z
\end{aligned} \tag{2.16}
$$

Consequently, the governing equations (or Navier equations of motion) in terms of rigidities can be expressed as

$$
\begin{aligned}
A_{11}\frac{\partial^2 u}{\partial x^2} - B_{11}\frac{\partial^3 w}{\partial x^3} + C_{11}\frac{\partial^2 v}{\partial x^2} &= \rho_0 \ddot{u} + \rho_{01}\ddot{v} - \rho_1 \frac{\partial \ddot{w}}{\partial x} \\
B_{11}\frac{\partial^3 u}{\partial x^3} - D_{11}\frac{\partial^4 w}{\partial x^4} + F_{11}\frac{\partial^3 v}{\partial x^3} &= \rho_1 \frac{\partial \ddot{u}}{\partial x} + \rho_{11}\frac{\partial \ddot{v}}{\partial x} + \rho_0 \ddot{w} - \rho_2 \frac{\partial^2 \ddot{w}}{\partial x^2} \\
C_{11}\frac{\partial^2 u}{\partial x^2} - F_{11}\frac{\partial^3 w}{\partial x^3} + E_{11}\frac{\partial^2 v}{\partial x^2} - H_{55} v &= \rho_{01}\ddot{u} + \rho_{02}\ddot{v} - \rho_{11}\frac{\partial \ddot{w}}{\partial x}
\end{aligned} \tag{2.17}
$$

Here, the extensional, coupling, and bending rigidities appearing in Eq. (2.17) are

$$A_{11} = \int_{-h/2}^{h/2} Q_{11}\, dz$$

$$(B_{11}, C_{11}) = \int_{-h/2}^{h/2} Q_{11}(z, f(z))\, dz$$

$$(D_{11}, F_{11}, E_{11}) = \int_{-h/2}^{h/2} Q_{11}(z^2, zf(z), f^2(z))\, dz$$

and transverse shear rigidity is

$$H_{55} = \int_{-h/2}^{h/2} Q_{55}[f'(z)]^2\, dz$$

Moreover, the notations of these rigidities may be different in numerical modeling based on the Rayleigh-Ritz method for vibration of FG beam.

2.4.2 Vibration of Thin FG Plate

Assuming the CPT given in Eq. (2.12), the free vibration problem associated with thin FG plate may be written as

$$\frac{\partial^2 M_x}{\partial x^2} - 2\frac{\partial^2 M_{xy}}{\partial x \partial y} + \frac{\partial^2 M_y}{\partial y^2} = I_0 \frac{\partial^2 w}{\partial t^2} \tag{2.18}$$

Here, bending moments are $M_x = -D_f\left(\frac{\partial^2 w}{\partial x^2} + \nu \frac{\partial^2 w}{\partial y^2}\right)$, $M_y = -D_f\left(\frac{\partial^2 w}{\partial y^2} + \nu \frac{\partial^2 w}{\partial x^2}\right)$ and twisting moment is $M_{xy} = D_f(1-\nu)\frac{\partial^2 w}{\partial x \partial y}$, where D_f is the flexural rigidity of the plate. $I_0 = \int_{-h/2}^{h/2} \rho(z)\, dz$ is the inertial coefficient.

Substituting the expressions of bending and twisting moments in Eq. (2.18), we can obtain the equation of equilibrium in terms of deflections (w) of the plate as

$$D_f\left(\frac{\partial^4 w}{\partial x^4} + 2\frac{\partial^4 w}{\partial x^2 \partial y^2} + \frac{\partial^4 w}{\partial y^4}\right) + I_0 \frac{\partial^2 w}{\partial t^2} = 0 \tag{2.19}$$

where

$$
\begin{aligned}
D_{\mathrm{f}} &= \int_{-h/2}^{h/2} \frac{z^2 E(z)}{1-\nu^2}\,\mathrm{d}z \\
&= \frac{1}{1-\nu^2}\int_{-h/2}^{h/2}\left\{(E_{\mathrm{c}}-E_{\mathrm{m}})\left(\frac{z}{h}+\frac{1}{2}\right)^k + E_{\mathrm{m}}\right\} z^2\,\mathrm{d}z \\
&= \frac{1}{1-\nu^2}\int_{-h/2}^{h/2}\left\{(E_{\mathrm{c}}-E_{\mathrm{m}})\left(\frac{z}{h}+\frac{1}{2}\right)^k\right\} z^2\,\mathrm{d}z + \int_{-h/2}^{h/2} E_{\mathrm{m}} z^2\,\mathrm{d}z \\
&= \frac{(E_{\mathrm{c}}-E_{\mathrm{m}})h^3}{1-\nu^2}\left\{\frac{1}{k+3}-\frac{1}{k+2}+\frac{1}{4(k+1)}\right\} + \frac{E_{\mathrm{m}}h^3}{12(1-\nu^2)}
\end{aligned}
$$

and

$$
\begin{aligned}
I_0 &= \int_{-h/2}^{h/2} \rho(z)\,\mathrm{d}z \\
&= \int_{-h/2}^{h/2}\left\{(\rho_{\mathrm{c}}-\rho_{\mathrm{m}})\left(\frac{z}{h}+\frac{1}{2}\right)^k + \rho_{\mathrm{m}}\right\}\mathrm{d}z \\
&= \int_{-h/2}^{h/2}\left\{(\rho_{\mathrm{c}}-\rho_{\mathrm{m}})\left(\frac{z}{h}+\frac{1}{2}\right)^k\right\}\mathrm{d}z + \int_{-h/2}^{h/2} \rho_{\mathrm{m}}\,\mathrm{d}z \\
&= \frac{(\rho_{\mathrm{c}}-\rho_{\mathrm{m}})h}{k+1} + \rho_{\mathrm{m}}h
\end{aligned}
$$

CHAPTER 3

Analytical and Numerical Methods

This chapter involves the higher-order partial differential equations concerned with the vibration problems of functionally graded (FG) beams and plates. The solution procedure may either be analytical or computational, depending upon the complexity of the partial differential equations (PDEs), as well as the boundary conditions. As such, firstly short reviews of various analytical and numerical methods are provided. Next, possible analytical method for solving vibration problem of FG beams and plates are incorporated; this is followed by the numerical procedure of the Rayleigh-Ritz method.

3.1 HISTORY OF VARIOUS METHODS

Identifying the vibration characteristics of FG beams and plates is the sole objective of the present investigation. In general, the analytical or closed form solution is the best possible solution to differential equation of any order. As such, various literature positions claim that different analytical and numerical techniques have been developed for these problems; these are surveyed below.

At first, analytical solution given to the vibration problems of isotropic as well as FG beams and plates are addressed. Reddy (1984) has given a refined nonlinear theory accounting for the von Karman strains to find exact solutions for simply supported plates. Gorman (1983, 1986, 1989) has proposed a highly accurate analytical solutions (method of superposition) for free vibration of right triangular plates with simply supported edge supports, with combinations of clamped-simply supported boundary supports and different boundary conditions with one edge free, respectively. Hosseini-Hashemi and Arsanjani (2005) have presented the exact closed form characteristic equations for the Mindlin plates with two opposite edges simply supported. The exact solutions for free vibration analysis of rectangular Mindlin plates have been obtained by Akhavan et al. (2009) under in-plane loads resting on a Pasternak elastic

Vibration of Functionally Graded Beams and Plates
http://dx.doi.org/10.1016/B978-0-12-804228-1.00003-1

foundation. A new beam theory was considered by Sina et al. (2009), different from traditional first-order shear deformation beam theory, to analyze the free vibration of FG beams through an analytical approach. Mahi et al. (2010) have presented exact solutions to study the free vibration of FG beam based on a unified higher-order shear deformation theory, in which material properties are temperature dependent. Using an analytical method, Thai and Vo (2012) have developed bending and free vibration of FG beams using various higher-order shear deformation beam theories. Exact closed-form solutions of 3D elasticity theory are presented in Hosseini-Hashemi et al. (2013) to study both in-plane and out-of-plane free vibrations for FG, simply supported rectangular plates.

Somehow, an analytical solution of the PDEs involved in certain complicated problems is not always possible. This fact allows the numerical solutions to be efficient in solving such problems, which certainly reduces the computational efforts. In general, a few of the widely used numerical methods are finite difference method (FDM), boundary element method (BEM), finite element method (FEM), differential quadrature method (DQM), collocation, Galerkin, Rayleigh-Ritz, meshless Petrov-Galerkin (MLPG), etc. In recent decades, a combination of methods, viz. mixed Ritz-DQM, mixed FEM-DQM, mixed FEM-Ritz, etc., have also been implemented in these studies. Other than these methods, the Rayleigh-Ritz method is one of the efficient numerical techniques that has received major attention for solving the problems of isotropic structural members. Different investigations performed based on this technique have been discussed here. In this regard, natural frequencies of simply supported elliptic plates are evaluated by Leissa (1967) by means of the Rayleigh-Ritz technique. The procedures followed by this method can easily be found in Timoshenko and Woinowsky-Krieger (1959), Kelly (1999), Rao (2004), and Chakraverty (2009). The characteristic orthogonal polynomials in the Rayleigh-Ritz method were first developed by Bhat (1986) to estimate the transverse vibration response of rotating cantilever beam with a tip mass. In another literature, Bhat (1987) has implemented a set of characteristic orthogonal polynomials to find the natural frequencies and mode shapes of polygonal plates by means of the Rayleigh-Ritz method. The natural frequencies of rectangular plates using characteristic orthogonal polynomials have been computed in Bhat (1985) and Cupial (1997) using the Rayleigh-Ritz method. Kim and Dickinson (1990, 1992) have studied free vibration of isotropic and orthotropic right triangular plates and of general triangular plates respectively by using the Rayleigh-Ritz method. Then Singh and

Chakraverty (1991, 1992a,b) have developed two-dimensional boundary characteristic orthogonal polynomials and studied transverse vibration of elliptic and circular plates using orthogonal polynomials in the Rayleigh-Ritz method satisfying different boundary conditions, viz. completely free, simply supported, and clamped, respectively. Axisymmetric vibration of circular and its analogous elliptic plates have been examined by Rajalingham and Bhat (1993) using characteristic orthogonal polynomials. Rajalingham et al. (1994) have analyzed vibration of clamped elliptic plates using exact circular plate modes as shape functions in the Rayleigh-Ritz method. Transverse vibrations of triangular plates have also been investigated by Singh and Chakraverty (1992c) with various types of boundary conditions at the edges by using boundary characteristics orthogonal polynomials as basis functions in the Rayleigh-Ritz method. The pb-2 Rayleigh-Ritz method is employed by Liew (1993) to study the free flexural vibration of triangular plates with/without curved internal supports. Free vibration analysis of thick cantilevered triangular plates is investigated by Karunasena et al. (1996) using the pb-2 Rayleigh-Ritz method based on the Mindlin shear deformation theory. Singh and Saxena (1996) have solved transverse vibration of triangular plates with variable thickness by working out several approximations in the Rayleigh-Ritz method. This method has also been employed by Singh and Hassan (1998) to obtain the numerical solution of a triangular plate with arbitrary thickness variation. A fast converging series consisting of a set of static beam functions has been developed by Ding (1996) to study vibration characteristics of thin rectangular plates. Ilanko (2009) has commented on the historical basis of the Rayleigh and Ritz method. Carrera et al. (2011) have implemented refined plate theories in finding accurate free vibration analysis of anisotropic simply supported plates. The flapwise and chordwise bending vibration analysis of rotating pre-twisted Timoshenko beam has been examined in Zhu (2011) by the use of the Rayleigh-Ritz method. Free vibration of the baffled circular plates with radial side cracks and in contact with water on one side has been studied by Si et al. (2012) based on the Rayleigh-Ritz method. Most recently, Pradhan and Chakraverty (2013, 2014a) have investigated free vibration of FG beams based on various shear deformation beam theories and have also proposed a new deformation beam theory in Pradhan and Chakraverty (2015). Moreover, Chakraverty and Pradhan (2014a,b) have studied free vibration of thin FG rectangular plates in the presence of complicating environments and that of thin FG elliptic plate in Pradhan and Chakraverty (2014b).

3.2 ANALYTICAL METHODS

This section includes very few available analytical solution procedures of handling vibration problems associated with FG beams and plates. In other words, one has to find the closed form solution of governing equations for FG beams and plates mentioned in Eqs. (2.17) and (2.19), respectively, satisfying the boundary conditions.

3.2.1 Vibration of the FG Beam

Let us first find the closed-form solution for equilibrium equations of simply supported FG beams. Accordingly, the following sets of boundary conditions at the edges of the beam are obtained at $x = 0$ and L (Aydogdu and Taskin, 2007):

$$N_x = w = M_x = M_x^f = 0$$

Navier solutions to Eq. (2.17) that satisfy the above boundary conditions can be expressed in the following form:

$$\begin{aligned} u &= A_{\mathrm{m}} \cos \frac{m\pi x}{L} \sin \omega t \\ v &= \frac{1}{L} B_{\mathrm{m}} \cos \frac{m\pi x}{L} \sin \omega t \\ w &= C_{\mathrm{m}} \sin \frac{m\pi x}{L} \sin \omega t \end{aligned} \tag{3.1}$$

where m is the half wave number in the x direction, ω is the natural frequency, and A_{m}, B_{m}, and C_{m} are undetermined constant coefficients. Inserting Eq. (3.1) in Eq. (2.17) leads to the following eigenvalue problem:

$$\left[\begin{pmatrix} K_{11} & K_{12} & K_{13} \\ K_{21} & K_{22} & K_{23} \\ K_{31} & K_{32} & K_{33} \end{pmatrix} - \omega^2 \begin{pmatrix} M_{11} & M_{12} & M_{13} \\ M_{21} & M_{22} & M_{23} \\ M_{31} & M_{32} & M_{33} \end{pmatrix} \right] \begin{Bmatrix} A_{\mathrm{m}} \\ B_{\mathrm{m}} \\ C_{\mathrm{m}} \end{Bmatrix} = 0. \tag{3.2}$$

Here,

$$K_{11} = A_{11}\frac{m^2\pi^2}{L^2}, \quad K_{12} = C_{11}\frac{m^2\pi^2}{L^3}, \quad K_{13} = B_{11}\frac{m^3\pi^3}{L^3}$$

$$K_{21} = B_{11}\frac{m^3\pi^3}{L^3}, \quad K_{22} = F_{11}\frac{m^3\pi^3}{L^4}, \quad K_{23} = D_{11}\frac{m^4\pi^4}{L^4}$$

$$K_{31} = C_{11}\frac{m^2\pi^2}{L^2}, \quad K_{32} = E_{11}\frac{m^2\pi^2}{L^3} + \frac{H_{55}}{L}, \quad K_{33} = F_{11}\frac{m^3\pi^3}{L^3}$$

$$M_{11} = \rho_0, \quad M_{12} = \rho_{01}, \quad M_{13} = \rho_1 \frac{m\pi}{L}$$

$$M_{21} = \rho_1 \frac{m\pi}{L}, \quad M_{22} = \rho_{11} \frac{m\pi}{L^2}, \quad M_{23} = \left(\rho_0 - \rho_2 \frac{m^2\pi^2}{L^2}\right)$$

$$M_{31} = \rho_{01}, \quad M_{32} = \rho_{02}, \quad M_{33} = \rho_{11} \frac{m\pi}{L}$$

In short matrix notation, we can write the eigenvalue problem (Eq. (3.2)) as $\left([K] - \omega^2[M]\right)\{\Delta\} = 0$ with $[K]$ and $[M]$ being stiffness and inertia matrices, respectively; Δ is the column vector of unknown coefficients. For different values of m, we can estimate a unique value for ω.

3.2.2 Vibration of the FG Plate

The analytical solutions for vibration of FG elliptic and triangular plates are sometimes different and these studies are scarce in open literature. But solutions of isotropic plates have already been studied rigorously using analytical and numerical methods. The detailed analysis can especially be found in Leissa (1969) and subsequently in Timoshenko and Woinowsky-Krieger (1959), Reddy (1993), Rao (2004, 2007), Chakraverty (2009), and Leissa and Qatu (2011).

The closed form solutions for equilibrium equations of simply supported FG rectangular plate (length a and breadth b with negligible thickness) will now be discussed and the following boundary conditions will be imposed at the edges; $x = 0,\ a$ and $y = 0,\ b$ (Thai et al., 2013):

$$w = M_x = M_y = 0$$

To solve the governing equation (2.19) meant for vibration of the FG plate, let us follow the Navier solution procedure. Assume the following solution form for w:

$$w = \sum_{m=1}^{\infty}\sum_{n=1}^{\infty} W_{mn} \exp(\mathrm{i}\omega t) \sin\alpha x \sin\beta y; \quad \text{with } \lambda = \frac{m\pi}{a},\ \beta = \frac{n\pi}{b} \tag{3.3}$$

where $\mathrm{i} = \sqrt{-1}$, ω is the natural frequency, and W_{mn} is unknown function to be determined.

Now, substituting Eq. (3.3) into Eq. (2.18) leads to the eigenvalue problem as below:

$$\left[D_{\mathrm{f}}\left(\alpha^2+\beta^2\right)^2-\omega^2 I_0\right]\{W_{mn}\}=0 \tag{3.4}$$

The natural frequencies of FG plate can be computed by setting the determinant of the coefficient matrix in Eq. (3.4) as zero. For each choice of m and n, there is a corresponding unique value of ω. The fundamental frequency is the smallest value of $\omega(m, n)$.

In both the vibration problems discussed in Sections 3.2.1 and 3.2.2, we have certainly assumed simply supported boundary conditions. This is due to the fact that generating Navier solution form for deflections under other classical edge supports, viz. clamped and free is very complicated and the analytical solution procedure for such edge conditions is very limited in open sources as well. In this regard, a trial and error solution-based technique (the so-called Lévy type procedure) is a widely used method in such findings.

3.3 NUMERICAL METHOD(S)

The Rayleigh-Ritz method (after Walther Ritz and Lord Rayleigh) is the numerical procedure implemented in the numerical modeling of different chapters. Vibration of uniform FG beams and thin FG plates will be discussed by means of the Rayleigh-Ritz method, as it was earlier extensively used in the analysis of isotropic structural members. However, the studies on FG structural members using this method is limited in open literature. In this regard, this book successfully implemented this computational procedure in handling vibration problems concerned with FG beams and plates. The summary of the algorithms for this technique is presented below.

1. The expressions for strain and kinetic energies of the FG beam (or plate) for the displacement field of the respective shear deformation beam (or plate) theories should be known.
2. The maximum strain and kinetic energies are obtained by assuming the harmonic type displacements.
3. We then consider the amplitudes of displacements as the linear combination of simple algebraic polynomials (generated from Pascal's triangle),

which also involves a specific admissible function to satisfy the essential boundary conditions.

4. Equate the maximum strain and kinetic energies to obtain the Rayleigh quotient. Then differentiate the Rayleigh quotient partially with respect to the unknown constant coefficients involved in linear combination to find the generalized eigenvalue problem.

3.3.1 Vibration of the FG Beam

Considering the constitutive relations related to the displacement field of higher-order shear deformation theory (Eq. (2.3)), the strain energy S and the kinetic energy T of the FG beam in cartesian coordinates may respectively be written as

$$S = \frac{1}{2}\int_{-L/2}^{L/2}\int_{A}(\sigma_{xx}\varepsilon_{xx} + \tau_{xz}\gamma_{xz})\,\mathrm{d}A\,\mathrm{d}x \tag{3.5}$$

and

$$T = \frac{1}{2}\int_{-L/2}^{L/2}\int_{A}\rho(z)\left[\left(\frac{\partial u_x}{\partial t}\right)^2 + \left(\frac{\partial u_z}{\partial t}\right)^2\right]\mathrm{d}A\,\mathrm{d}x \tag{3.6}$$

where A and ρ are the area of cross-section and the mass density of the rectangular FG beam, respectively. Considering the higher-order shear deformation beam theory mentioned in Eq. (2.4), Eqs. (3.5) and (3.6) become

$$S = \frac{1}{2}\int_{-L/2}^{L/2}\left[A_{xx}u_{,x}^2 - 2B_{xx}u_{,x}w_{,xx} + D_{xx}w_{,xx}^2 + 2E_{xx}u_{,x}v_{,x} - 2F_{xx}v_{,x}w_{,xx} + H_{xx}v_{,x}^2 + A_{xz}v^2\right]\mathrm{d}x \tag{3.7}$$

and

$$T = \frac{1}{2}\int_{-L/2}^{L/2}\left[I_A\left\{\left(\frac{\partial u}{\partial t}\right)^2 + \left(\frac{\partial w}{\partial t}\right)^2\right\} - 2I_B\left(\frac{\partial u}{\partial t}\right)\left(\frac{\partial^2 w}{\partial x\partial t}\right) + I_D\left(\frac{\partial^2 w}{\partial x\partial t}\right)^2 + 2I_E\left(\frac{\partial u}{\partial t}\right)\left(\frac{\partial v}{\partial t}\right) - 2I_F\left(\frac{\partial v}{\partial t}\right)\left(\frac{\partial^2 w}{\partial x\partial t}\right) + I_H\left(\frac{\partial v}{\partial t}\right)^2\right]\mathrm{d}x \tag{3.8}$$

The stiffness coefficients appearing in Eq. (3.7) are defined as

$$
\begin{aligned}
(A_{xx}, B_{xx}, D_{xx}) &= \int_A Q_{11}(1, z, z^2)\,\mathrm{d}A \\
(E_{xx}, F_{xx}) &= \int_A f(z) Q_{11}(1, z)\,\mathrm{d}A \\
H_{xx} &= \int_A [f(z)]^2 Q_{11}\,\mathrm{d}A \\
A_{xz} &= \kappa \int_A [f'(z)]^2 Q_{55}\,\mathrm{d}A \quad (\kappa\text{-shear correction factor})
\end{aligned}
\tag{3.9}
$$

In Eq. (3.9), $\kappa = 5/6$ is introduced in the case of TBT, whereas it will be unity for all other SDBTs. As we are neglecting both transverse shear and transverse normal effects in the case of CBT, it can be seen that terms associated with $f(z)$ and shear correction factor play no role while evaluating the expression of strain energy. The cross-sectional inertial coefficients in Eq. (3.8) may be written in the following form:

$$
\begin{aligned}
(I_A, I_B, I_D) &= \int_A \rho(z)(1, z, z^2)\,\mathrm{d}A \\
(I_E, I_F) &= \int_A f(z)\rho(z)(1, z)\,\mathrm{d}A \\
I_H &= \int_A [f(z)]^2 \rho(z)\,\mathrm{d}A
\end{aligned}
\tag{3.10}
$$

Assuming the displacement components $u(x, t)$, $v(x, t)$, and $w(x, t)$ as the harmonic type, those may be expressed as

$$
\begin{aligned}
u(x, t) &= U(x) \sin \omega t \\
v(x, t) &= V(x) \sin \omega t \\
w(x, t) &= W(x) \sin \omega t
\end{aligned}
\tag{3.11}
$$

where $U(x)$, $V(x)$, and $W(x)$ are the respective amplitudes for these displacement components of free vibration of FG beam and the trigonometric terms indicate the harmonic type displacement with ω as the natural frequency. Substituting Eq. (3.11) into Eqs. (3.7) and (3.8) leads to maximum strain energy ($S_{\max}$) and maximum kinetic energy ($T_{\max}$) as

$$S_{\max} = \frac{1}{2}\int_{-L/2}^{L/2}\left[A_{xx}\left(\frac{\partial U}{\partial x}\right)^2 - 2B_{xx}\left(\frac{\partial U}{\partial x}\right)\left(\frac{\partial^2 W}{\partial x^2}\right) + D_{xx}\left(\frac{\partial^2 W}{\partial x^2}\right)^2 \right.$$
$$+ 2E_{xx}\left(\frac{\partial U}{\partial x}\right)\left(\frac{\partial V}{\partial x}\right) - 2F_{xx}\left(\frac{\partial V}{\partial x}\right)\left(\frac{\partial^2 W}{\partial x^2}\right)$$
$$\left. + H_{xx}\left(\frac{\partial V}{\partial x}\right)^2 + A_{xz}V^2\right]\mathrm{d}x \qquad (3.12)$$

and

$$T_{\max} = \frac{\omega^2}{2}\int_{-L/2}^{L/2}\left[I_A\left(U^2 + W^2\right) - 2I_B U\frac{\partial W}{\partial x} + I_D\left(\frac{\partial W}{\partial t}\right)^2 \right.$$
$$\left. + 2I_E UV - 2I_F V\frac{\partial W}{\partial x} + I_H V^2\right]\mathrm{d}x \qquad (3.13)$$

For the Rayleigh-Ritz method, the amplitudes of vibration are expressed in terms of algebraic polynomial functions by the following series (Timoshenko and Woinowsky-Krieger, 1959; Kelly, 1999; Rao, 2004; Chakraverty, 2009):

$$U = \sum_{i=1}^{n} c_i\varphi_i, \quad V = \sum_{j=1}^{n} d_j\psi_j, \quad W = \sum_{k=1}^{n} e_k f_k$$

where c_i, d_j, and e_k are the unknown constant coefficients to be determined and φ_i, ψ_j, and f_k are the admissible functions, which must satisfy the essential boundary conditions and can be represented as

$$\varphi_i = fx^{i-1}, \quad i = 1, 2, \ldots, n$$

$$\psi_j = fx^{j-1}, \quad j = 1, 2, \ldots, n$$

$$f_k = fx^{k-1}, \quad k = 1, 2, \ldots, n$$

Here, n is the number of polynomials involved in the admissible functions (Chakraverty, 2009; Bhat, 1986; Singh and Chakraverty, 1992a,b) and $f = \left(x + \frac{L}{2}\right)^p \left(x - \frac{L}{2}\right)^q$ where p, $q = 0, 1$ or 2 are given in Table 3.1, as per the six classical boundary conditions (combination of clamped (C), simply supported (S), and free (F)) of rectangular FG beam.

Accordingly, the Rayleigh quotient (ω^2) can be found by equating $S_{\max}$ and $T_{\max}$ obtained in Eqs. (3.12) and (3.13). We take the partial derivative

Table 3.1 Admissible function indices for different boundary conditions of rectangular FG beam

BCs	p	q
C-C	2	2
C-S	2	1
C-F	2	0
S-S	1	1
S-F	1	0
F-F	0	0

Source: Reprinted from International Journal of Mechanical Sciences, Vol 82, Pradhan, K.K., Chakraverty, S., Effects of different shear deformation theories on free vibration of functionally graded beams, Pages 149–160, Copyright (2014), with permission from Elsevier.

of this quotient with respect to the constant coefficients involved in the admissible functions as

$$\frac{\partial \omega^2}{\partial c_i} = 0; \quad i = 1, 2, \ldots, n$$
$$\frac{\partial \omega^2}{\partial d_j} = 0; \quad j = 1, 2, \ldots, n$$
$$\frac{\partial \omega^2}{\partial e_k} = 0; \quad k = 1, 2, \ldots, n$$

As such, we may obtain the generalized eigenvalue problem for free vibration of FG beam as

$$\left([K] - \lambda^2[M]\right)\{\Delta\} = 0 \tag{3.14}$$

where $[K]$ and $[M]$ are the stiffness and inertia matrices, respectively, and $\{\Delta\}$ is the column vector of unknown coefficients. The eigenvalue (λ) for the above eigenvalue problem (Eq. (3.14)) is the non-dimensional frequency and its different non-dimensional expressions have been given in its respective chapter.

3.3.2 Vibration of the FG Plate

Considering the constitutive relations associated with classical plate theory stated in Eq. (2.14), the strain energy **U** and kinetic energy **T** of the plate in cartesian coordinates may be written as

$$\mathbf{U} = \frac{1}{2}\int_{\Omega}\left[\int_{-h/2}^{h/2}\left(\sigma_{xx}\epsilon_{xx} + \sigma_{yy}\epsilon_{yy} + \tau_{xy}\gamma_{xy}\right)\,\mathrm{d}z\right]\mathrm{d}x\,\mathrm{d}y \tag{3.15}$$

and

$$\mathbf{T}=\frac{1}{2}\int_{\Omega}\left[\int_{-h/2}^{h/2}\rho(z)\left(\frac{\partial u_z}{\partial t}\right)^2\mathrm{d}z\right]\mathrm{d}x\,\mathrm{d}y \tag{3.16}$$

where Ω denotes the midplane (domain) of the FG plate of any geometry viz. rectangular, elliptic, or triangular. Using Eqs. (2.12), (2.13), and (2.14) in Eqs. (3.15) and (3.16) leads to

$$\mathbf{U}=\frac{1}{2}\int_{\Omega}\left[D_{11}\left\{\left(\frac{\partial^2 w}{\partial x^2}\right)^2+\left(\frac{\partial^2 w}{\partial y^2}\right)^2\right\}+2D_{12}\frac{\partial^2 w}{\partial x^2}\frac{\partial^2 w}{\partial y^2}\right.$$
$$\left.+\,4D_{66}\left(\frac{\partial^2 w}{\partial x\partial y}\right)^2\right]\mathrm{d}x\,\mathrm{d}y \tag{3.17}$$

and

$$\mathbf{T}=\frac{1}{2}\int_{\Omega}I_0\left(\frac{\partial w}{\partial t}\right)^2\mathrm{d}x\,\mathrm{d}y \tag{3.18}$$

where the stiffness coefficients in Eq. (3.17) are

$$(D_{11},D_{12},D_{66})=\int_{-h/2}^{h/2}(Q_{11},Q_{12},Q_{66})z^2\,\mathrm{d}z$$

and the inertial coefficient, I_0 in Eq. (3.18), is

$$I_0=\int_{-h/2}^{h/2}\rho(z)\,\mathrm{d}z$$

The displacement component can be assumed to be the harmonic type as $w(x,y,t) = W(x,y)\cos\omega t$ with $W(x,y)$ as the maximum deflection and ω is the natural frequency of free vibration. Using the above harmonic deflection, Eqs. (3.17) and (3.18) may be transformed into maximum strain ($\mathbf{U}_{\max}$) and kinetic energies ($\mathbf{T}_{\max}$), respectively, as follows:

$$\mathbf{U}_{\max}=\frac{1}{2}\int_{\Omega}\left[D_{11}\left\{\left(\frac{\partial^2 W}{\partial x^2}\right)^2+\left(\frac{\partial^2 W}{\partial y^2}\right)^2\right\}+2D_{12}\frac{\partial^2 W}{\partial x^2}\frac{\partial^2 W}{\partial y^2}\right.$$
$$\left.+\,4D_{66}\left(\frac{\partial^2 W}{\partial x\partial y}\right)^2\right]\mathrm{d}x\,\mathrm{d}y \tag{3.19}$$

and

$$\mathbf{T}_{\max} = \frac{\omega^2}{2} \int_{\Omega} I_0 W^2 \, dx \, dy \tag{3.20}$$

In the Rayleigh-Ritz method, one may express the transverse displacement ($W(x, y)$) as the sum of simple algebraic polynomials involving both x and y (as done before):

$$W(x, y) = \sum_{i=1}^{n} c_i \varphi_i(x, y) \tag{3.21}$$

where c_i are unknown constants to be determined and φ_i are the admissible functions, which satisfy the essential boundary conditions and can be represented as

$$\varphi_i(x, y) = f \psi_i(x, y), \quad i = 0, 1, 2, \ldots, n \tag{3.22}$$

Here, n is the number of polynomials involved in the admissible functions. The function f in Eq. (3.22) is generally chosen by the geometry of the plate, viz. rectangular, elliptic, or triangular, as given below:

$$\begin{aligned} &\text{Rectangular}: f = x^p y^q (a - x)^r (b - y)^s \\ &\text{Elliptic}: f = \left(1 - \frac{x^2}{a^2} - \frac{y^2}{b^2}\right)^p \\ &\text{Triangular}: f = x^p y^q (1 - x - y)^r \end{aligned} \tag{3.23}$$

where the indices p, q, r, and s decide the classical boundary conditions at the edges. The parameter p takes the value as 0, 1, or 2 according the edge as free, simply supported, or clamped, respectively. Similar interpretations can also be given to other parameters. Table 3.2 involves the components of ψ_i generated from Pascal's triangle.

Assuming constant Poisson's ratio (ν), the Rayleigh quotient can be obtained by equating $\mathbf{U}_{\max}$ and $\mathbf{T}_{\max}$ as

Table 3.2 Ten algebraic polynomials obtained from Pascal's triangle

i	1	2	3	4	5	6	7	8	9	10
ψ_i	1	x	y	x^2	xy	y^2	x^3	x^2y	xy^2	y^3

Source: Reproduced from Chakraverty, S., Pradhan, K.K., Free vibration of exponential functionally graded rectangular plates in thermal environment with general boundary conditions. Aerospace Science and Technology 2014;36:132–156.

$$\omega^2 = \frac{\int_\Omega D_{11}\left[\left\{\left(\frac{\partial^2 W}{\partial x^2}\right)^2 + \left(\frac{\partial^2 W}{\partial y^2}\right)^2\right\} + 2\nu\frac{\partial^2 W}{\partial x^2}\frac{\partial^2 W}{\partial y^2} + 2(1-\nu)\left(\frac{\partial^2 W}{\partial x \partial y}\right)^2\right] \mathrm{d}x\,\mathrm{d}y}{\int_\Omega I_0 W^2 \,\mathrm{d}x\,\mathrm{d}y} \tag{3.24}$$

Now we take the partial derivative of ω^2 with respect to unknown constants as

$$\frac{\partial \omega^2}{\partial c_i} = 0, \quad i = 1, 2, 3, \ldots, n \tag{3.25}$$

Further manipulation of Eq. (3.25) yields the generalized eigenvalue problem of the form

$$\left([K]_{n\times n} - \lambda^2 [M]_{n\times n}\right)\{\Delta\} = 0 \tag{3.26}$$

where $[K]_{n\times n}$ and $[M]_{n\times n}$ are symmetric stiffness and inertia matrices, respectively, and $\{\Delta\}$ is the column vector of unknown constant coefficients. Solutions of the eigenvalue problem, Eq. (3.26), give the vibration characteristics, viz. non-dimensional frequencies and mode shapes for free vibration of FG plate with different geometric configurations. On the other hand, the numerical modeling for triangular FG plate needs minor modifications due to the transformation of geometry from the general cartesian coordinate system to standard triangle. In similar fashion, change in expression of strain energy of deformed FG plate occurs in the presence of different complicating environments. Therefore, the details about their modeling have been discussed in the relevant chapter.

CHAPTER 4

Functionally Graded Beams

The vibration problems concerned with isotropic structural members and various composites have been studied rigorously in earlier decades. But the research efforts on vibration analysis of functionally graded (FG) beams is limited. On the other hand, these problems are to be solved using different shear deformation beam theories (SDBTs) to estimate vibration characteristics of structural beams, and the applied methods may not handle all sets of classical boundary conditions as well. A few of the significant investigations associated with vibration of FG beams are discussed below.

As such, the free vibration behavior of a simply supported FG beam has been studied by Aydogdu and Taskin (2007) by using the Euler-Bernoulli beam theory, parabolic shear deformation theory, and exponential shear deformation theory. A new beam theory was considered by Sina et al. (2009) that was different from traditional first-order SDBT, to investigate the free vibration of FG beams through an analytical approach. Şimşek (2010b) has examined the vibration response of a simply supported FG beam to a moving mass by using Euler-Bernoulli, Timoshenko, and the third-order SDBTs. Using different higher-order SDBTs, Şimşek (2010a) has also recently studied the fundamental frequencies of FG beams subjected to different boundary conditions. Mahi et al. (2010) have presented exact solutions to study the free vibration of FG beams based on a unified higher-order shear deformation theory, in which material properties are temperature dependent. An improved third order shear deformation theory is introduced in Wattanasakulpong et al. (2011) to study thermal buckling and elastic vibration of FG beams. Alshorbagy et al. (2011) have used the finite element method to study the free vibration characteristics of a FG beam. Shahba et al. (2011) have investigated free vibration and stability analysis of axially FG Timoshenko tapered beams using classical and non-classical boundary conditions through the finite element approach. Using an analytical method, Thai and Vo (2012) have developed bending and free vibration of FG beams employing various higher-order SDBTs. Free vibration and

http://dx.doi.org/10.1016/B978-0-12-804228-1.00004-3

stability of axially FG tapered Euler-Bernoulli beams are investigated by Shahba and Rajasekaran (2012) using the finite element method. Size-dependent linear free flexural vibration behavior of FG nanoplates is investigated in Natarajan et al. (2012) using the iso-geometric-based finite element method. Şimşek (2012) has found the free longitudinal vibration of axially FG tapered nanorods based on the nonlocal elasticity theory. Free vibration analysis of size-dependent FG nanobeams is studied by Eltaher et al. (2012) using the finite element method. Static and buckling behaviors of nonlocal FG Timoshenko nanobeam are investigated in Eltaher et al. (2014) based on the finite element method. Vo et al. (2013) have presented static and vibration analysis of FG beams using refined shear deformation theory by using the finite element formulation. Static and free vibration of axially loaded rectangular FG beams is developed in Nguyen et al. (2013) based on the first-order SDBT. The plane stress problem of an orthotropic FG beam with arbitrary graded material properties along the thickness direction is investigated recently by the displacement function approach in Nie et al. (2013). Huang et al. (2013) have investigated the vibration behavior of axially FG Timoshenko beams with non-uniform cross-section by introducing an auxiliary function in the coupled governing equations. Also, the study of dynamic characteristics of FG beams based on different shear deformation theories may have been discussed in Kahrobaiyan et al. (2012), Lei et al. (2013), Rahmani and Pedram (2014), Kien (2014), Komijani et al. (2014), Sharma et al. (2014) and the articles provided therein.

This chapter deals with free vibration of FG beams based on various SDBTs. Here a few previously proposed deformation theories are considered and, in addition, alternate forms of newly proposed deformation theories are also included, based on certain assumptions. The material properties in FG beams vary continuously along thickness direction in power-law form. The numerical procedures of the Rayleigh-Ritz method (as stated in Chapter 3) are implemented here to obtain the generalized eigenvalue problem. Comparison of non-dimensional frequencies is addressed after the test of convergence to report the corresponding results.

4.1 VIBRATION OF UNIFORM FG BEAM

As mentioned earlier, the aim of this section is to investigate free vibration of uniform FG beam by means of the Rayleigh-Ritz method. As such, new results for natural frequencies are to be computed by finding the generalized eigenvalue problem for the problem concerned.

4.1.1 Numerical Modeling

The mathematical modeling for free vibration of FG beams has already been discussed in Section 3.3.1 to generate the eigenvalue problem (Eq. (3.14)). The comparison of present results with available literature has been carried out after checking the test of convergence. As such, the non-dimensional frequencies of FG beam under different sets of boundary conditions can be obtained based on all the SDBTs as mentioned in Eqs. (2.4) and (2.5). The results based on SDBTs of Eq. (2.4) have already been obtained by Pradhan and Chakraverty (2014a), and are included in this chapter. It is easy to compute the natural frequencies of all the newly proposed beam theories of Eq. (2.5) and exhaustive results for these theories are not included here. Moreover, a few results associated with PESDBT of Eq. (2.5) are reported here, but detailed results can be found in Pradhan and Chakraverty (2015).

4.1.2 Convergence and Comparison Studies

In this section, the convergence of first five non-dimensional frequencies of FG beams subjected to different sets of boundary conditions has been performed. The material properties of FG beam constituents are given below (Sina et al., 2009).

The subscripted terms in Table 4.1, that is, $(\ )_{\mathrm{m}}$ and $(\ )_{\mathrm{c}}$, are the material properties of the metal and ceramic constituents of FG beam, respectively. As mentioned in Sina et al. (2009), the non-dimensional frequencies will be evaluated as follows:

$$\lambda = \frac{\omega L^2}{h}\sqrt{\frac{I_A}{\int_{-h/2}^{h/2} E(z)\,\mathrm{d}z}}, \quad I_A = \int_{-h/2}^{h/2} \rho(z)\,\mathrm{d}z \tag{4.1}$$

In Tables 4.2 and 4.3, the convergence of first five non-dimensional frequencies for C-C FG beams is checked with the increase in the

Table 4.1 Material properties of the FG beam constituents (aluminum and Al_2O_3)

Properties	Unit	Aluminum; $(\ )_{\mathrm{m}}$	Alumina; $(\ )_{\mathrm{c}}$
E	GPa	70	380
ρ	$\mathrm{kg/m^3}$	2700	3800
ν	-	0.23	0.23

Source: Reprinted from International Journal of Mechanical Sciences, Vol 82, Pradhan, K.K., Chakraverty, S., Effects of different shear deformation theories on free vibration of functionally graded beams, Pages 149–160, Copyright (2014), with permission from Elsevier.

Table 4.2 Convergence of the first five non-dimensional frequencies of C-C FG beams with $L/h = 5$ and $k = 1$ using CBT

n	λ_1	λ_2	λ_3	λ_4	λ_5
2	6.0184	15.2859	18.7637	33.1976	-
5	5.8682	15.0208	16.3766	27.4765	32.7191
8	5.8286	15.0160	16.1670	27.3988	32.1442
11	5.8190	15.0131	16.0119	27.3740	31.9793
14	5.8110	15.0121	15.9643	27.3548	31.8450
15	5.8105	15.0115	15.9404	27.3537	31.8376

Source: Reprinted from International Journal of Mechanical Sciences, Vol 82, Pradhan, K.K., Chakraverty, S., Effects of different shear deformation theories on free vibration of functionally graded beams, Pages 149–160, Copyright (2014), with permission from Elsevier.

Table 4.3 Convergence of the first five non-dimensional frequencies of C-C FG beams with $L/h = 5$ and $k = 1$ using TBT

n	λ_1	λ_2	λ_3	λ_4	λ_5
2	5.5977	15.2791	18.6079	32.6268	57.1052
5	5.2016	12.5717	16.1091	21.6487	32.0403
8	5.1219	12.1873	15.9422	21.0416	30.4953
11	5.0748	12.0647	15.8154	20.5341	29.9598
14	5.0590	11.9534	15.7824	20.3872	29.4853
15	5.0515	11.9507	15.7610	20.2969	29.4774

Source: Reprinted from International Journal of Mechanical Sciences, Vol 82, Pradhan, K.K., Chakraverty, S., Effects of different shear deformation theories on free vibration of functionally graded beams, Pages 149–160, Copyright (2014), with permission from Elsevier.

Table 4.4 Convergence of the first five non-dimensional frequencies of S-S FG beams with $L/h = 5$ and $k = 1$ using CBT

n	λ_1	λ_2	λ_3	λ_4	λ_5
2	3.1109	11.2824	17.4533	32.4072	-
5	2.7550	9.5117	15.8728	20.4816	31.8019
8	2.7550	9.4798	15.8286	20.1998	31.5139
9	2.7550	9.4798	15.8286	20.1957	31.4957
10	2.7550	9.4798	15.8286	20.1957	31.4957

Source: Reprinted from International Journal of Mechanical Sciences, Vol 82, Pradhan, K.K., Chakraverty, S., Effects of different shear deformation theories on free vibration of functionally graded beams, Pages 149–160, Copyright (2014), with permission from Elsevier.

number of polynomials in displacement components using CBT and TBT, respectively. In similar fashion, the convergence pattern of these frequencies of S-S FG beam are reported in Tables 4.4 and 4.5 using CBT and TBT, respectively. In these tables, slenderness ratio (L/h) is taken as 5 and power-law index (k) as unity. Moreover, first five non-dimensional

Table 4.5 Convergence of the first five non-dimensional frequencies of S-S FG beams with $L/h = 5$ and $k = 1$ using TBT

n	λ_1	λ_2	λ_3	λ_4	λ_5
2	3.1108	11.0467	17.1864	31.7900	53.0719
5	2.6513	8.6317	15.4036	17.2822	30.5461
8	2.6468	8.6042	15.3925	17.0444	26.4038
9	2.6450	8.6030	15.3920	16.9850	26.4032
10	2.6449	8.5928	15.3890	16.9820	26.2741

Source: Reprinted from International Journal of Mechanical Sciences, Vol 82, Pradhan, K.K., Chakraverty, S., Effects of different shear deformation theories on free vibration of functionally graded beams, Pages 149–160, Copyright (2014), with permission from Elsevier.

Table 4.6 Convergence of the first five non-dimensional frequencies of S-S FG beams with $k = 0.3$ using TBT for different L/h

L/h	n	λ_1	λ_2	λ_3	λ_4	λ_5
10	3	2.7436	13.5064	31.5140	34.0368	65.0296
	5	2.7396	10.3895	22.4190	31.2411	62.5095
	8	2.7382	10.3736	22.0642	31.1597	36.7517
	10	2.7378	10.3679	22.0229	31.1592	36.6360
	Sina et al. (2009)	2.695	-	-	-	-
	Şimşek (2010a)	2.701	-	-	-	-
30	3	2.7756	13.8995	36.4061	94.2770	194.4858
	5	2.7744	10.9368	24.7787	74.4866	94.9697
	8	2.7742	10.8982	24.3695	42.8160	70.2439
	10	2.7742	10.8974	24.3570	42.7468	66.2351
	Sina et al. (2009)	2.737	-	-	-	-
	Şimşek (2010a)	2.738	-	-	-	-
100	3	2.7793	13.9457	36.7130	314.1700	648.0928
	5	2.7784	11.0100	25.1099	76.5300	141.0338
	8	2.7784	10.9643	24.6941	43.8368	73.8106
	10	2.7784	10.9642	24.6874	43.7776	68.6585
	Sina et al. (2009)	2.742	-	-	-	-
	Şimşek (2010a)	2.742	-	-	-	-

Source: Reprinted from International Journal of Mechanical Sciences, Vol 82, Pradhan, K.K., Chakraverty, S., Effects of different shear deformation theories on free vibration of functionally graded beams, Pages 149–160, Copyright (2014), with permission from Elsevier.

frequencies of S-S and C-C FG beams based on TBT are incorporated in Tables 4.6 and 4.7, respectively, with different slenderness ratios and $k = 0.3$. The fundamental frequencies are also validated with Sina et al. (2009) and Şimşek (2010a) in Tables 4.6 and 4.7, which are found to be

Table 4.7 Convergence of the first five non-dimensional frequencies of C-C FG beams with $k = 0.3$ using TBT for different L/h

L/h	n	λ_1	λ_2	λ_3	λ_4	λ_5
10	3	6.0025	16.4281	32.9760	34.2325	66.3549
	8	5.9085	15.3645	28.2359	31.9735	43.4821
	13	5.8874	15.2536	27.8470	31.6367	42.7630
	17	5.8808	15.2117	27.7302	31.5568	42.4941
	Sina et al. (2009)	5.811	-	-	-	-
	Şimşek (2010a)	5.875	-	-	-	-
30	3	6.2168	17.4307	35.6795	98.9047	199.0017
	8	6.1853	16.9236	32.8346	53.6489	79.6691
	13	6.1807	16.8955	32.7663	53.4353	78.5944
	17	6.1788	16.8877	32.7400	53.3791	78.4686
	Sina et al. (2009)	6.167	-	-	-	-
	Şimşek (2010a)	6.177	-	-	-	-
100	3	6.2428	17.5693	35.8678	329.6765	663.3261
	8	6.2201	17.1370	33.5584	55.4290	83.8482
	13	6.2171	17.1236	33.5370	55.3615	82.5777
	17	6.2159	17.1203	33.5284	55.3477	82.5496
	Sina et al. (2009)	6.212	-	-	-	-
	Şimşek (2010a)	6.214	-	-	-	-

in good agreement. It can be observed that an increase in the number of polynomials plays a key role in the convergence of non-dimensional frequencies. On the contrary, one may also verify the convergence of non-dimensional frequencies of FG beams under any set of edge supports based on other SDBTs.

For the shear deformation theories, viz. parabolic shear deformation beam theory (PSDBT), exponential shear deformation beam theory (ESDBT), trigonometric shear deformation beam theory (TSDBT), a new shear deformation beam theory (ASDBT), power-law exponent based shear deformation beam theory (PESDBT), and inverse trigonometric SDBTs, the physical properties of FG constituents are taken as set out in Thai and Vo (2012), Vo et al. (2013), and Şimşek (2010a): $E_m = 70$ GPa, $\rho_m = 2702\,\text{kg/m}^3$, $E_c = 380$ GPa, $\rho_c = 3960\,\text{kg/m}^3$ and $\nu_m = \nu_c = 0.3$ with different slenderness ratios (L/h) and power-law indices (k). For the convergence test in the newly proposed PESDBT, the notation for number

Table 4.8 Convergence of the first five non-dimensional frequencies of FG beams with $L/h = 5$ and $k = 1$ using PSDBT

BCs	n	λ_1	λ_2	λ_3	λ_4	λ_5
C-C	2	8.9371	24.6923	29.9774	52.5714	84.4010
	3	8.5034	22.6687	27.1270	47.6067	55.7833
	4	8.3502	20.0462	26.4320	43.6627	52.8966
	6	8.1878	19.4048	25.9525	33.3363	48.2137
	9	8.0522	19.0412	25.5961	32.2934	47.2373
	10	8.0441	18.8861	25.5918	32.2532	46.5707
S-S	2	5.0230	17.8230	27.7230	51.3506	83.1496
	3	4.2928	17.7191	27.5078	41.1175	56.3212
	4	4.2801	13.9833	24.9005	40.0057	54.5206
	6	4.2723	13.8818	24.8416	27.3943	43.4178
	9	4.2647	13.8502	24.8290	27.3134	42.4286
	10	4.2646	13.8396	24.8259	27.3105	42.2756

Source: Reprinted from Applied Mathematics and Computation, Vol 268, Pradhan, K.K., Chakraverty, S., Generalized power-law exponent based shear deformation theory for free vibration of functionally graded beams, Pages 1240–1258, Copyright (2015), with permission from Elsevier.

Table 4.9 Convergence of the first five non-dimensional frequencies of FG beams with $L/h = 10$ and $k = 0.2$ using ESDBT

BCs	n	λ_1	λ_2	λ_3	λ_4	λ_5
C-C	2	11.2275	32.0043	63.7141	121.899	371.647
	3	11.1202	30.3946	60.6124	63.2878	121.963
	4	11.0304	29.0417	59.5323	60.5496	106.869
	6	10.9853	28.6427	52.4646	59.3656	81.3672
	9	10.9440	28.3878	51.9135	58.4866	79.9793
	10	10.9384	28.3186	51.8177	58.4855	79.4583
S-S	2	5.6850	25.1861	58.2237	119.097	370.155
	3	5.0677	25.1472	57.8426	63.2132	119.344
	4	5.0604	19.4014	57.6009	62.6609	115.735
	6	5.0593	19.2880	41.0265	57.5336	69.9765
	9	5.0571	19.2648	40.8079	57.5214	67.8157
	10	5.0570	19.2581	40.8001	57.5210	67.6176

of polynomials is m, and n (power-law index denoting PESDBT) is taken as unity. In Tables 4.8–4.15, the non-dimensional frequency may be computed by the following formulation to check the convergence:

$$\lambda = \frac{\omega L^2}{h}\sqrt{\frac{\rho_m}{E_m}} \tag{4.2}$$

Table 4.10 Convergence of the first five non-dimensional frequencies of FG beams with $L/h = 10$ and $k = 0.2$ using TSDBT

BCs	n	λ_1	λ_2	λ_3	λ_4	λ_5
C-C	2	11.2259	32.0039	63.7140	121.899	371.282
	3	11.1183	30.3875	60.6123	63.2851	121.963
	4	11.0282	29.0311	59.5140	60.5495	106.863
	6	10.9829	28.6303	52.4293	59.3655	81.2894
	9	10.9413	28.3740	51.8734	58.4866	79.8969
	10	10.9357	28.3039	51.7773	58.4855	79.3704
S-S	2	5.6850	25.1849	58.2237	119.097	369.788
	3	5.0675	25.1457	57.8426	63.2035	119.344
	4	5.0602	19.3978	57.6007	62.6490	115.735
	6	5.0590	19.2841	41.0060	57.5331	69.9227
	9	5.0568	19.2604	40.7885	57.5209	67.7663
	10	5.0566	19.2535	40.7805	57.5205	67.5647

Table 4.11 Convergence of the first five non-dimensional frequencies of FG beams with $L/h = 5$ and $k = 1$ using ASDBT

BCs	n	λ_1	λ_2	λ_3	λ_4	λ_5
C-C	2	8.9422	24.6924	29.9792	52.5777	84.3990
	3	8.5111	22.6854	27.1303	47.6068	55.7929
	4	8.3580	20.0849	26.4339	43.6945	52.9073
	6	8.1971	19.4510	25.9538	33.4480	48.4590
	9	8.0645	19.0940	25.5973	32.4362	47.5044
	10	8.0564	18.9458	25.5929	32.3956	46.8624
S-S	2	5.0230	17.8257	27.7256	51.3565	83.1504
	3	4.2941	17.7224	27.5106	41.1313	56.3470
	4	4.2815	13.9968	24.9067	40.0259	54.5470
	6	4.2741	13.8981	24.8480	27.4683	43.6172
	9	4.2671	13.8695	24.8364	27.3938	42.6105
	10	4.2671	13.8610	24.8339	27.3913	42.4764

Source: Reprinted from Applied Mathematics and Computation, Vol 268, Pradhan, K.K., Chakraverty, S., Generalized power-law exponent based shear deformation theory for free vibration of functionally graded beams, Pages 1240–1258, Copyright (2015), with permission from Elsevier.

It can easily be said that the non-dimensional frequencies gradually converge with an increase in the number of polynomials regardless of SDBT and other parameters considered.

After looking into the satisfactory test of convergence, one may compare the non-dimensional frequencies of FG beams associated with different edge supports. In Tables 4.16 and 4.17, fundamental frequencies with

Table 4.12 Convergence of the first five non-dimensional frequencies of FG beams with $L/h = 5$, $k = 1$, and $n = 1$ using PESDBT

BCs	n	λ_1	λ_2	λ_3	λ_4	λ_5
C-C	2	8.9371	24.6923	29.9774	52.5714	84.4010
	3	8.5034	22.6687	27.1270	47.6067	55.7833
	4	8.3502	20.0462	26.4320	43.6627	52.8966
	6	8.1878	19.4048	25.9525	33.3363	48.2137
	9	8.0522	19.0412	25.5961	32.2934	47.2373
	10	8.0441	18.8861	25.5918	32.2532	46.5707
S-S	2	5.0230	17.8230	27.7230	51.3506	83.1496
	3	4.2928	17.7191	27.5078	41.1175	56.3212
	4	4.2801	13.9833	24.9005	40.0057	54.5206
	6	4.2723	13.8818	24.8416	27.3943	43.4178
	9	4.2647	13.8502	24.8290	27.3134	42.4286
	10	4.2646	13.8396	24.8259	27.3105	42.2756

Source: Reprinted from Applied Mathematics and Computation, Vol 268, Pradhan, K.K., Chakraverty, S., Generalized power-law exponent based shear deformation theory for free vibration of functionally graded beams, Pages 1240–1258, Copyright (2015), with permission from Elsevier.

Table 4.13 Convergence of the first five non-dimensional frequencies of FG beams with $L/h = 10$ and $k = 0.2$ using ISDBT

BCs	n	λ_1	λ_2	λ_3	λ_4	λ_5
C-C	2	11.1613	31.9817	63.7125	121.890	340.409
	3	11.0402	30.1140	60.6111	63.1673	121.953
	4	10.9397	28.6225	58.8674	60.5479	106.595
	6	10.8864	28.1665	51.2309	59.3642	78.8402
	9	10.8360	27.8802	50.5719	58.4853	77.4177
	10	10.8303	27.7940	50.4774	58.4844	76.8166
S-S	2	5.6850	25.1287	58.2221	119.089	338.881
	3	5.0579	25.0837	57.8410	62.7869	119.329
	4	5.0497	19.2548	57.5934	62.1758	115.721
	6	5.0482	19.1359	40.3511	57.5157	68.2576
	9	5.0454	19.1083	40.1620	57.4999	66.2398
	10	5.0452	19.0995	40.1538	57.4991	66.0231

Table 4.14 Convergence of the first five non-dimensional frequencies of FG beams with $L/h = 10$ and $k = 0.2$ using ICDBT

BCs	n	λ_1	λ_2	λ_3	λ_4	λ_5
C-C	2	11.1613	31.9817	63.7125	121.890	340.408
	3	11.0402	30.1140	60.6111	63.1672	121.953
	4	10.9396	28.6225	58.8674	60.5479	106.595
	6	10.8864	28.1665	51.2310	59.3642	78.8404
	9	10.8360	27.8802	50.5720	58.4853	77.4180
	10	10.8303	27.7941	50.4775	58.4844	76.8169

Continued

Table 4.14 Convergence of the first five non-dimensional frequencies of FG beams with $L/h = 10$ and $k = 0.2$ using ICDBT—cont'd

BCs	n	λ_1	λ_2	λ_3	λ_4	λ_5
S-S	2	5.6850	25.1288	58.2222	119.089	338.341
	3	5.0580	25.0839	57.8411	62.7885	119.329
	4	5.0498	19.2560	57.5935	62.1792	115.721
	6	5.0483	19.1376	40.3642	57.5160	68.2950
	9	5.0455	19.1107	40.1749	57.5002	66.2744
	10	5.0454	19.1025	40.1671	57.4995	66.0648

Table 4.15 Convergence of the first five non-dimensional frequencies of FG beams with $L/h = 10$ and $k = 0.2$ using ITDBT

BCs	n	λ_1	λ_2	λ_3	λ_4	λ_5
C-C	2	11.1617	31.9818	63.7126	121.891	339.868
	3	11.0406	30.1165	60.6112	63.1676	121.953
	4	10.9402	28.6263	58.8757	60.5480	106.596
	6	10.8871	28.1716	51.2493	59.3643	78.8902
	9	10.8370	27.8867	50.5966	58.4853	77.4755
	10	10.8313	27.8019	50.5024	58.4844	76.8827
S-S	2	5.6850	25.1288	58.2222	119.089	338.341
	3	5.0580	25.0839	57.8411	62.7885	119.329
	4	5.0498	19.2560	57.5935	62.1792	115.721
	6	5.0483	19.1376	40.3642	57.5160	68.2950
	9	5.0455	19.1107	40.1749	57.5002	66.2744
	10	5.0454	19.1025	40.1671	57.4995	66.0648

Table 4.16 Comparison of non-dimensional fundamental frequencies of FG beams with $k = 0$

			L/h		
BCs	SDBT	Source	10	30	100
S-S	CBT	Present	2.837	2.847	2.849
		Sina et al. (2009)	2.849	2.849	2.849
		Şimşek (2010b)	2.837	2.847	2.848
	TBT	Present	2.805	2.844	2.848
		Sina et al. (2009)	2.797	2.843	2.848
		Şimşek (2010b)	2.804	2.843	2.848
	PSDBT	Present	2.803	2.844	2.849
	ESDBT	Present	2.804	2.844	2.848
	HSDBT	Present	2.803	2.844	2.849
	TSDBT	Present	2.818	2.846	2.849
	ASDBT	Present	2.804	2.844	2.849

Source: Reprinted from International Journal of Mechanical Sciences, Vol 82, Pradhan, K.K., Chakraverty, S., Effects of different shear deformation theories on free vibration of functionally graded beams, Pages 149–160, Copyright (2014), with permission from Elsevier.

Table 4.17 Comparison of non-dimensional fundamental frequencies of FG beams with $k = 0.3$

BCs	SDBT	Source	L/h 10	30	100
S-S	CBT	Present	2.768	2.778	2.779
		Şimşek (2010b)	2.731	2.741	2.743
	TBT	Present	2.738	2.774	2.778
		Sina et al. (2009)	2.695	2.737	2.742
		Şimşek (2010b)	2.701	2.738	2.742
		Şimşek (2010a)	2.701	2.738	2.742
	PSDBT	Present	2.736	2.774	2.778
		Şimşek (2010a)	2.702	2.738	2.742
	ESDBT	Present	2.737	2.774	2.778
	HSDBT	Present	2.737	2.774	2.778
	TSDBT	Present	2.749	2.775	2.779
	ASDBT	Present	2.736	2.774	2.778
		Şimşek (2010a)	2.702	2.738	2.742
C-F	CBT	Present	0.975	0.977	0.977
	TBT	Present	0.971	0.977	0.977
		Sina et al. (2009)	0.969	0.976	0.977
		Şimşek (2010a)	0.970	0.976	0.977
	PSDBT	Present	0.970	0.976	0.977
		Şimşek (2010a)	0.970	0.976	0.977
	ESDBT	Present	0.970	0.977	0.977
	HSDBT	Present	0.970	0.976	0.977
	TSDBT	Present	0.972	0.977	0.977
	ASDBT	Present	0.970	0.976	0.977
		Şimşek (2010a)	0.970	0.976	0.977
C-C	CBT	Present	6.188	6.216	6.220
	TBT	Present	5.881	6.179	6.216
		Sina et al. (2009)	5.811	6.167	6.212
		Şimşek (2010a)	5.875	6.177	6.214
	PSDBT	Present	5.874	6.178	6.215
		Şimşek (2010a)	5.881	6.177	6.214
	ESDBT	Present	5.873	6.177	6.215
	HSDBT	Present	5.871	6.177	6.215
	TSDBT	Present	5.993	6.193	6.215
	ASDBT	Present	5.874	6.178	6.216
		Şimşek (2010a)	5.884	6.177	6.214

Source: Reprinted from International Journal of Mechanical Sciences, Vol 82, Pradhan, K.K., Chakraverty, S., Effects of different shear deformation theories on free vibration of functionally graded beams, Pages 149–160, Copyright (2014), with permission from Elsevier.

different slenderness ratios ($L/h = 10, 30, 100$) and a fixed k are compared with the study of Sina et al. (2009) and Şimşek (2010a,b) based on all the SDBTs using the expression given in Eq. (4.1). FG constituents own the properties (Sina et al., 2009): $E_m = 70$ GPa, $\rho_m = 2700\,kg/m^3$, $E_c = 380$ GPa, $\rho_c = 3800\,kg/m^3$, and $\nu_m = \nu_c = 0.23$. In Table 4.16, the non-dimensional fundamental frequencies of S-S FG beams with $k = 0$ within the framework of all the SDBTs have been evaluated, and comparisons are performed with the results for CBT and FSDBT^2[1] (TBT) as provided in Sina et al. (2009) and with those for CBT and TBT as mentioned in Şimşek (2010b). In a similar fashion, non-dimensional fundamental frequencies of S-S, C-F, and C-C FG beams with $k = 0.3$ using all the SDBTs are computed in Table 4.17 and comparisons are carried out with the results for FSDBT^2 (TBT) of Sina et al. (2009), with those obtained using CBT and TBT as given in Şimşek (2010b) and with the non-dimensional fundamental frequencies using FSDBT^S[2] (TBT), PSDBT^S[3] (PSDBT), and ASDBT^S[4] (ASDBT), as considered in Şimşek (2010a). One may clearly notice here that the fundamental non-dimensional frequencies obtained in the present investigation are approximately close enough to the results provided in these literatures that are used for comparison.

For the verification of present results in Tables 4.18–4.20, natural frequencies are computed on the basis of the formulation as stated in Eq. (4.2). In Table 4.18, the properties of the FG constituents are assumed to be $E_m = 70$ GPa, $E_c = 380$ GPa, and $\nu_m = \nu_c = 0.3$, as mentioned in Aydogdu and Taskin (2007). Considering no gradation of mass density through the thickness of FG constituents (constant density), non-dimensional fundamental frequencies of S-S FG beam are compared with Aydogdu and Taskin (2007) in Table 4.18 based on CBT, TBT, PSDBT, and ESDBT. Physical properties of FG constituents in Tables 4.19 and 4.20 are taken as set out in Şimşek (2010a), Thai and Vo (2012), and Vo et al. (2013): $E_m = 70$ GPa, $\rho_m = 2702\,kg/m^3$, $E_c = 380$ GPa, $\rho_c = 3960\,kg/m^3$, and $\nu_m = \nu_c = 0.3$. Assuming no role of Poisson's ratio (ν) in the expression of Q_{11} (reduced stiffness coefficient), comparison of fundamental frequencies is carried out in Table 4.19 with Şimşek

[1] FSDBT^2 in Sina et al. (2009) is the same as TBT in the present investigation.
[2] FSDBT^S in Şimşek (2010a) is the same as TBT in the present study.
[3] PSDBT^S in Şimşek (2010a) is the same as PSDBT in the present study.
[4] ASDBT^S in Şimşek (2010a) is the same as ASDBT in the present study.

Table 4.18 Comparison of non-dimensional fundamental frequencies of S-S FG beams with Aydogdu and Taskin (2007) for different slenderness ratios (L/h)

L/h	Theory	Source	$k = 0$	$k = 0.1$	$k = 1$	$k = 2$	$k = 10$
5	CBT	Present	6.847	6.512	5.176	4.752	3.959
		Aydogdu and Taskin (2007)	6.847	6.499	4.821	4.251	3.737
	TBT	Present	6.569	6.254	4.974	4.549	3.756
		Aydogdu and Taskin (2007)	6.563	6.237	4.652	4.101	3.563
	PSDBT	Present	6.526	6.215	4.942	4.506	3.685
		Aydogdu and Taskin (2007)	6.574	6.248	4.659	4.103	3.548
	ESDBT	Present	6.531	6.220	4.945	4.509	3.688
		Aydogdu and Taskin (2007)	6.584	6.258	4.665	4.109	3.553
20	CBT	Present	6.951	6.612	5.256	4.826	4.021
		Aydogdu and Taskin (2007)	6.951	6.599	4.907	4.334	3.804
	TBT	Present	6.932	6.593	5.242	4.811	4.006
		Aydogdu and Taskin (2007)	6.931	6.580	4.895	4.323	3.791
	PSDBT	Present	6.928	6.589	5.239	4.807	3.999
		Aydogdu and Taskin (2007)	6.932	6.581	4.895	4.323	3.790
	ESDBT	Present	6.928	6.589	5.239	4.808	3.999
		Aydogdu and Taskin (2007)	6.933	6.582	4.896	4.323	3.790

Source: Reprinted from International Journal of Mechanical Sciences, Vol 82, Pradhan, K.K., Chakraverty, S., Effects of different shear deformation theories on free vibration of functionally graded beams, Pages 149—160, Copyright (2014), with permission from Elsevier.

(2010a), Thai and Vo (2012), and Vo et al. (2013) subjected to various BCs such as C-C, C-F, and S-S, within the framework of CBT and TBT. In a similar way, TBT is considered in Table 4.20 for comparison of non-dimensional higher-order frequencies for S-S and C-C FG beams with Thai and Vo (2012), Vo et al. (2013), and Nguyen et al. (2013). A very good agreement of present results can be observed for fundamental and also for higher-order natural frequencies of FG beams irrespective of the edge supports. A little discrepancy for third non-dimensional frequencies can be observed in Table 4.20 for S-S FG beams with $L/h = 5$. But it can be viewed that fourth frequencies for different power-law exponents (k) are

Table 4.19 Comparison of non-dimensional fundamental frequencies of FG beams for different k with various BCs

L/h	Theory	BC	Source	$k = 0$	$k = 0.2$	$k = 0.5$	$k = 1$	$k = 2$	$k = 5$	$k = 10$
5	CBT	C-C	Present	12.1826	11.3410	10.3875	9.3993	8.5772	8.1515	7.9041
			Şimşek (2010a)	12.1826	11.3398	10.3718	9.36422	8.52772	8.10955	7.87968
		C-F	Present	1.9385	1.8042	1.6524	1.4954	1.3655	1.2989	1.2593
			Şimşek (2010a)	1.93845	1.8042	1.65057	1.49135	1.35985	1.29416	1.25648
		S-S	Present	5.3953	5.0538	4.7314	4.4488	4.2177	3.9617	3.6977
			Şimşek (2010a)	5.39533	5.02194	4.59360	4.14835	3.77930	3.59487	3.49208
			Thai and Vo (2012)	5.3953	-	4.5936	4.1484	3.7793	3.5949	3.4921
	TBT	C-C	Present	10.0456	9.4267	8.7121	7.9401	7.2281	6.6826	6.3525
			Şimşek (2010a)	10.0344	9.41764	8.70047	7.92529	7.21134	6.66764	6.34062
			Vo et al. (2013)	9.99836	9.38337	-	7.90153	7.19013	6.64465	6.31609
		C-F	Present	1.9021	1.7717	1.6233	1.4688	1.3396	1.2706	1.2301
			Şimşek (2010a)	1.89479	1.76554	1.61737	1.46300	1.33376	1.26445	1.22398
			Vo et al. (2013)	1.89442	1.76477	-	1.46279	1.33357	1.26423	1.22372
		S-S	Present	5.1546	4.8364	4.5313	4.2566	4.0198	3.7464	3.4886
			Şimşek (2010a)	5.15247	4.80657	4.40830	3.99023	3.63438	3.43119	3.31343
			Thai and Vo (2012)	5.1527	-	4.4107	3.9904	3.6264	3.4012	3.2816
			Vo et al. (2013)	5.15260	4.80328	-	3.97108	3.60495	3.40253	3.29625
20	CBT	C-C	Present	12.4142	11.5549	10.5871	9.5907	8.7684	8.3425	8.0797
			Şimşek (2010a)	12.4142	11.5537	10.5713	9.55538	8.71856	8.30064	8.05560
		C-F	Present	1.9525	1.8171	1.6644	1.5070	1.3771	1.3105	1.2699
			Şimşek (2010a)	1.95248	1.81714	1.66265	1.50293	1.37142	1.30574	1.26713

	S-S	Present	5.4777	5.1299	4.8024	4.5161	4.2832	4.0252	3.7568
		Şimşek (2010a)	5.47773	5.09804	4.66458	4.21634	3.84719	3.66283	3.55465
		Thai and Vo (2012)	5.4777	-	4.6641	4.2163	3.8472	3.6628	3.5547
TBT	C-C	Present	12.2252	11.3850	10.4320	9.4435	8.6203	8.1838	7.9215
		Şimşek (2010a)	12.2235	11.3850	10.4263	9.43135	8.60401	8.16985	7.91275
		Vo et al. (2013)	12.2202	11.3795	-	9.43114	8.60467	8.16977	7.91154
	C-F	Present	1.9501	1.8147	1.6612	1.5025	1.3715	1.3054	1.2661
		Şimşek (2010a)	1.94957	1.81456	1.66044	1.50104	1.36968	1.30375	1.26495
		Vo et al. (2013)	1.94955	1.81408	-	1.50106	1.36970	1.30376	1.26495
	S-S	Present	5.4605	5.1144	4.7881	4.5023	4.2690	4.0095	3.7415
		Şimşek (2010a)	5.46032	5.08265	4.65137	4.20505	3.83676	3.65088	3.54156
		Thai and Vo (2012)	5.4603	-	4.6516	4.2050	3.8361	3.6485	3.5390
		Vo et al. (2013)	5.46033	5.08120	-	4.20387	3.83491	3.64903	3.54045

Source: Reprinted from International Journal of Mechanical Sciences, Vol 82, Pradhan, K.K., Chakraverty, S., Effects of different shear deformation theories on free vibration of functionally graded beams, Pages 149–160, Copyright (2014), with permission from Elsevier.

Table 4.20 Comparison of non-dimensional natural frequencies of FG beams for higher modes with various BCs based on TBT

BC	L/h	Mode	Source	$k=0$	$k=0.2$	$k=0.5$	$k=1$	$k=2$	$k=5$	$k=10$
S-S	5	2	Present	17.8908	16.7317	15.3306	13.7778	12.3619	11.4822	11.1126
			Thai and Vo (2012)	17.8812	-	15.4588	14.0100	12.6405	11.5431	11.0240
			Nguyen et al. (2013)	18.5019	17.3654	16.0161	14.5160	13.0562	11.8698	11.3436
		3	Present (4th mode)	34.2103	32.1672	29.7504	27.0657	24.4881	22.4290	21.3110
			Present	30.2314	28.8311	27.0804	24.9042	22.3517	19.5600	18.0553
			Thai and Vo (2012)	34.2097	-	29.8382	27.0979	24.3152	21.7158	20.5561
			Nguyen et al. (2013)	35.0951	33.11059	30.6771	27.8565	24.8641	22.0568	20.9045
	20	2	Present	21.5755	20.0852	18.3849	16.6146	15.1402	14.3804	13.9476
			Thai and Vo (2012)	21.5732	-	18.3962	16.6344	15.1619	14.3746	13.9263
			Nguyen et al. (2013)	22.5873	21.0309	19.2616	17.4189	15.8723	15.0404	14.5721
		3	Present	47.6037	44.3818	40.7639	37.0394	33.9159	32.0700	30.8925
			Thai and Vo (2012)	47.5930	-	40.6526	36.7679	33.4689	31.5780	30.5369
			Nguyen et al. (2013)	49.7603	46.3777	42.5121	38.4544	34.9818	32.9705	31.8869
C-C	5	2	Present	23.1004	21.7700	20.2077	18.4654	16.7457	15.2301	14.3617
			Vo et al. (2013)	23.87540	22.48400	-	19.04940	17.29240	15.78680	14.90350
		3	Present	30.3513	28.9837	27.3569	25.3859	22.9576	19.9122	18.2304
			Vo et al. (2013)	30.23910	28.88370	-	25.37460	23.01120	19.96340	18.23210
		4	Present	38.6867	36.5446	34.0075	31.1243	28.1712	25.4176	23.8724
			Vo et al. (2013)	38.1841	36.0793	-	30.7500	27.8331	25.0901	23.5501
	20	2	Present	33.0067	30.7650	28.2156	25.5579	23.3181	22.0692	21.3200
			Vo et al. (2013)	33.1335	30.8452	-	25.6223	23.3691	22.1345	21.4015
		3	Present	63.0338	58.8118	53.9936	48.9397	44.6208	42.0801	40.5636
			Vo et al. (2013)	62.9124	58.7017	-	48.8401	44.5197	41.9748	40.4612
		4	Present	100.9961	94.3357	86.7036	78.6406	71.6395	67.2872	64.7111
			Vo et al. (2013)	101.2440	94.6356	-	78.8259	71.5625	66.5576	63.9421

Source: Reprinted from International Journal of Mechanical Sciences, Vol 82, Pradhan, K.K., Chakraverty, S., Effects of different shear deformation theories on free vibration of functionally graded beams, Pages 149–160, Copyright (2014), with permission from Elsevier.

in excellent agreement with that of Thai and Vo (2012) and Nguyen et al. (2013).

On the other hand, the fundamental non-dimensional frequencies of FG beams based on different SDBTs, viz. PSDBT, ESDBT, TSDBT, ASDBT, PESDBT, ISDBT, ICDBT, and ITDBT, are validated in Tables 4.21–4.23 with available results for different L/h and k. C-C edge support is considered in Table 4.21 and in similar fashion, Tables 4.22 and 4.23 are meant for C-F (cantilever) and S-S edge conditions, respectively. The present results are again found to be in excellent agreement with existing ones for the mentioned SDBTs.

4.1.3 Results and Discussions

In view of the above acceptable validation of non-dimensional fundamental and higher-order frequencies, the first five non-dimensional frequencies of FG beam are reported with different sets of BCs taking all SDBTs. The new results have been evaluated based on previously proposed SDBTs (given in Eq. (2.4)) in Tables 4.24–4.29 only for three sets of BCs, viz. C-C, S-S, and C-F, with different slenderness ratios ($L/h = 5$ and 20) and power-law indices ($k = 0, 0.1, 1, 2, 10$). The constituent material properties in these computations are stated in Table 4.1.

To report the present results, Poisson's ratio (ν) will play its role in the expression of Q_{11} and it will remain constant through the thickness of the FG beam. Other FG constituents such as Young's modulus and mass densities are assumed to vary along the thickness as per the power-law exponent form. In Tables 4.24 and 4.25, the effect of change of power-law indices on the first five non-dimensional frequencies of C-C FG beams are reported with $L/h = 5$ and $L/h = 20$, respectively. In similar fashion, the effect of variation of power-law exponents on these frequencies are evaluated with S-S edge support in Tables 4.26 and 4.27 and with C-F edge support in Tables 4.28 and 4.29. The number of polynomials (n) involved in displacement components for C-C FG beam is approximately 17, whereas it is 20 for S-S and 25 for C-F FG beams, respectively.

One may easily conclude that non-dimensional frequencies are increasing with an increase in slenderness ratios (L/h) and are decreasing with an increase in power-law exponents (k). It is also interesting to note that the non-dimensional frequencies for $L/h = 5$ using CBT are comparatively greater than those found using other SDBTs, whereas one may experience mere coincidence of frequencies for $L/h = 20$ while comparing the results obtained for different SDBTs of Eq. (2.4). So looking into the new results

Table 4.21 Comparison of fundamental non-dimensional frequencies of C-C FG beams based on different SDBTs with $L/h = 5$ and 20

L/h	Theory	Source	$k = 0$	$k = 0.2$	$k = 0.5$	$k = 1$	$k = 2$	$k = 5$	$k = 10$
5	PSDBT	Present	10.0858	9.4789	8.7614	7.9680	7.1963	6.5120	6.1809
		Şimşek (2010a)	10.0705	9.46641	8.74674	7.95034	7.17674	6.49349	6.16515
	ESDBT	Present	10.1099	9.5001	8.7802	7.9848	7.2082	6.5105	6.1918
		Şimşek (2010a)	10.0944	9.48737	8.76530	7.96695	7.18839	6.49186	6.17594
	TSDBT	Present	10.1099	9.5001	8.7802	7.9848	7.2082	6.5105	6.1918
		Şimşek (2010a)	10.0797	9.47451	8.75394	7.95676	7.18062	6.49028	6.16794
	ASDBT	Present	10.1099	9.5001	8.7802	7.9848	7.2082	6.5105	6.1918
		Şimşek (2010a)	10.0944	9.48737	8.76530	7.96695	7.18839	6.49186	6.17594
	PESDBT	$n = 1$	10.0858	9.4789	8.7614	7.9680	7.1963	6.5120	6.1809
		$n = 5$	10.1767	9.5581	8.8294	8.0314	7.2837	6.6822	6.3258
		$n = 10$	10.2373	9.6093	8.8136	8.0735	7.3335	6.7596	6.4115
	ISDBT	Present	9.7797	9.2030	8.5177	7.7534	6.9967	6.3003	5.9591
	ICDBT	Present	9.7796	9.2031	8.5178	7.7534	6.9967	6.3003	5.9591
	ITDBT	Present	9.7966	9.2179	8.5311	7.7653	7.0030	6.2911	5.9620
20	PSDBT	Present	12.2256	11.3873	10.4344	9.4438	8.6138	8.1587	7.8947
		Şimşek (2010a)	12.2238	11.3873	10.4287	9.43158	8.59751	8.14460	7.88576
	ESDBT	Present	12.2269	11.3884	10.4354	9.4446	8.6141	8.1569	7.8945
		Şimşek (2010a)	12.2251	11.3884	10.4297	9.43242	8.59775	8.14280	7.88558
	TSDBT	Present	12.2269	11.3884	10.4354	9.4446	8.6141	8.1569	7.8945
		Şimşek (2010a)	12.2242	11.3876	10.4290	9.43182	8.59739	8.14346	7.88534
	ASDBT	Present	12.2269	11.3884	10.4354	9.4446	8.6141	8.1569	7.8945
		Şimşek (2010a)	12.2251	11.3884	10.4297	9.43242	8.59775	8.14280	7.88558
	PESDBT	$n = 1$	12.2256	11.3873	10.4344	9.4438	8.6138	8.1587	7.8947
		$n = 5$	12.2396	11.3991	10.4443	9.4529	8.6260	8.1834	7.9178
		$n = 10$	12.2464	11.4047	10.4491	9.4574	8.6314	8.1925	7.9285
	ISDBT	Present	12.1891	11.3554	10.4069	9.4199	8.5909	8.1306	7.8634
	ICDBT	Present	12.1891	11.3554	10.4069	9.4199	8.5909	8.1306	7.8634
	ITDBT	Present	12.1894	11.3556	10.4072	9.4201	8.5903	8.1274	7.8618

Source: Reprinted from Applied Mathematics and Computation, Vol 268, Pradhan, K.K., Chakraverty, S., Generalized power-law exponent based shear deformation theory for free vibration of functionally graded beams, Pages 1240–1258, Copyright (2015), with permission from Elsevier.

Table 4.22 Comparison of fundamental non-dimensional frequencies of cantilever FG beams based on different SDBTs with $L/h = 5$ and 20

L/h	Theory	Source	$k = 0$	$k = 0.2$	$k = 0.5$	$k = 1$	$k = 2$	$k = 5$	$k = 10$
5	PSDBT	Present	1.8955	1.7663	1.6187	1.4645	1.3341	1.2605	1.2192
		Şimşek (2010a)	1.89523	1.76637	1.61817	1.46328	1.33254	1.25916	1.21834
	ESDBT	Present	1.8959	1.7666	1.6190	1.4647	1.3342	1.2602	1.2193
		Şimşek (2010a)	1.89565	1.76673	1.61848	1.46357	1.33268	1.25889	1.21843
	TSDBT	Present	1.8956	1.7664	1.6188	1.4646	1.3341	1.2603	1.2192
		Şimşek (2010a)	1.89536	1.76649	1.61827	1.46339	1.33256	1.25896	1.21831
	ASDBT	Present	1.8955	1.7663	1.6187	1.4645	1.3341	1.2605	1.2192
		Şimşek (2010a)	1.89565	1.76673	1.61848	1.46357	1.33268	1.25889	1.21843
	PESDBT	$n = 1$	1.8955	1.7663	1.6187	1.4645	1.3341	1.2605	1.2192
		$n = 5$	1.8983	1.7687	1.6207	1.4664	1.3366	1.2658	1.2241
		$n = 10$	1.8999	1.7700	1.6218	1.4674	1.3378	1.2678	1.2265
	ISDBT	Present	1.8872	1.7590	1.6125	1.4591	1.3289	1.2543	1.2123
	ICDBT	Present	1.8875	1.7593	1.6130	1.4598	1.3299	1.2551	1.2128
	ITDBT	Present	1.8874	1.7592	1.6126	1.4592	1.3289	1.2537	1.2121
20	PSDBT	Present	1.9496	1.8143	1.6609	1.5021	1.3711	1.3046	1.2653
		Şimşek (2010a)	1.94954	1.81458	1.66049	1.50106	1.36957	1.30332	1.26453
	ESDBT	Present	1.9496	1.8143	1.6609	1.5022	1.3711	1.3046	1.2653
		Şimşek (2010a)	1.94960	1.81464	1.66049	1.50106	1.36957	1.30332	1.26453
	TSDBT	Present	1.9496	1.8143	1.6609	1.5022	1.3711	1.3046	1.2653
		Şimşek (2010a)	1.94960	1.81458	1.66049	1.50106	1.36957	1.30332	1.26453
	ASDBT	Present	1.9496	1.8143	1.6609	1.5022	1.3711	1.3046	1.2653
		Şimşek (2010a)	1.94960	1.81464	1.66049	1.50106	1.36957	1.30332	1.26453
	PESDBT	$n = 1$	1.9496	1.8143	1.6609	1.5021	1.3711	1.3046	1.2653
		$n = 5$	1.9498	1.8145	1.6610	1.5023	1.3713	1.3050	1.2656
		$n = 10$	1.9499	1.8146	1.6611	1.5024	1.3714	1.3052	1.2658
	ISDBT	Present	1.9490	1.8138	1.6604	1.5018	1.3707	1.3042	1.2648
	ICDBT	Present	1.9490	1.8138	1.6606	1.5022	1.3713	1.3047	1.2651
	ITDBT	Present	1.9490	1.8138	1.6604	1.5018	1.3707	1.3041	1.2647

Source: Reprinted from Applied Mathematics and Computation, Vol 268, Pradhan, K.K., Chakraverty, S., Generalized power-law exponent based shear deformation theory for free vibration of functionally graded beams, Pages 1240–1258, Copyright (2015), with permission from Elsevier.

Table 4.23 Comparison of fundamental non-dimensional frequencies of S-S FG beams based on different SDBTs with $L/h = 5$ and 20

L/h	Theory	Source	$k = 0$	$k = 0.2$	$k = 0.5$	$k = 1$	$k = 2$	$k = 5$	$k = 10$
5	PSDBT	Present	5.1629	4.8459	4.5405	4.2632	4.0165	3.7178	3.4617
		Şimşek (2010a)	5.15274	4.80924	4.41108	3.99042	3.62643	3.40120	3.28160
	ESDBT	Present	5.1665	4.8492	4.5435	4.2661	4.0189	3.7179	3.4640
		Şimşek (2010a)	5.15422	4.81053	4.41222	3.99139	3.62671	3.39905	3.28134
		Thai and Vo (2012)	5.1542	-	4.4118	3.9914	3.6267	3.3991	3.2814
	TSDBT	Present	5.1643	4.8472	4.5417	4.2643	4.0174	3.7174	3.4624
		Şimşek (2010a)	5.15422	4.81053	4.41222	3.99139	3.62671	3.39905	3.28134
		Thai and Vo (2012)	5.1531	-	4.4110	3.9907	3.6263	3.3998	3.2811
	ASDBT	Present	5.1665	4.8492	4.5435	4.2661	4.0189	3.7179	3.4640
		Şimşek (2010a)	5.15422	4.81053	4.41222	3.99139	3.62671	3.39905	3.28134
	PESDBT	$n = 1$	5.1629	4.8459	4.5405	4.2632	4.0165	3.7178	3.4617
		$n = 5$	5.1729	4.8546	4.5483	4.2712	4.0299	3.7469	3.4846
		$n = 10$	5.1807	4.8612	4.5542	4.2775	4.0383	3.7603	3.4986
	ISDBT	Present	5.1199	4.8073	4.5050	4.2288	3.9802	3.6766	3.4209
	ICDBT	Present	5.1199	4.8074	4.5050	4.2288	3.9802	3.6766	3.4209
	ITDBT	Present	5.1228	4.8099	4.5074	4.2311	3.9817	3.6752	3.4219
20	PSDBT	Present	5.4606	5.1147	4.7884	4.5024	4.2682	4.0064	3.7386
		Şimşek (2010a)	5.46030	5.08286	4.65159	4.20503	3.83611	3.64850	3.53896

ESDBT	Present	5.4607	5.1148	4.7885	4.5025	4.2682	4.0062	3.7386
	Şimşek (2010a)	5.46042	5.08292	4.65165	4.20515	3.83617	3.64832	3.53896
	Thai and Vo (2012)	5.4604	-	4.6512	4.2051	3.8361	3.6483	3.5390
TSDBT	Present	5.4606	5.1147	4.7884	4.5024	4.2682	4.0063	3.7386
	Şimşek (2010a)	5.46036	5.08286	4.65159	4.20509	3.83611	3.64838	3.53896
	Thai and Vo (2012)	5.4603	-	4.6511	4.2051	3.8361	3.6484	3.5389
ASDBT	Present	5.4607	5.1148	4.7885	4.5025	4.2682	4.0062	3.7386
	Şimşek (2010a)	5.46042	5.08292	4.65165	4.20515	3.83617	3.64832	3.53896
PESDBT	$n = 1$	5.4606	5.1147	4.7884	4.5024	4.2682	4.0064	3.7386
	$n = 5$	5.4618	5.1157	4.7893	4.5034	4.2697	4.0094	3.7411
	$n = 10$	5.4624	5.1163	4.7898	4.5039	4.2704	4.0105	3.7423
ISDBT	Present	5.4572	5.1116	4.7856	4.4997	4.2652	4.0029	3.7351
ICDBT	Present	5.4572	5.1116	4.7856	4.4997	4.2652	4.0029	3.7351
ITDBT	Present	5.4572	5.1116	4.7856	4.4997	4.2652	4.0026	3.7350

Source: Reprinted from Applied Mathematics and Computation, Vol 268, Pradhan, K.K., Chakraverty, S., Generalized power-law exponent based shear deformation theory for free vibration of functionally graded beams, Pages 1240–1258, Copyright (2015), with permission from Elsevier.

Table 4.24 First five non-dimensional frequencies of C-C beams with $L/h = 5$ and different power-law indices (k)

Theory	k	λ_1	λ_2	λ_3	λ_4	λ_5
	0	12.7709	31.8774	33.4380	61.0565	63.7983
	0.1	12.3056	31.1160	32.2394	58.8549	62.3138
CBT	1	9.8372	25.4355	26.9706	46.3706	53.8664
	2	8.9681	22.9435	24.5910	41.6450	49.0744
	10	8.2747	19.1082	21.5967	37.3798	39.9629
	0	10.3768	23.7563	31.8427	39.5711	56.6681
	0.1	10.0436	23.0453	31.0890	38.4343	55.0875
TBT	1	8.2185	19.0273	26.6176	31.9040	45.9129
	2	7.4822	17.2580	24.0597	28.8844	41.5295
	10	6.5540	14.7564	19.1090	24.3983	34.7827
	0	10.4297	24.0206	31.8774	40.2276	57.8710
	0.1	10.1054	23.3343	31.1231	39.1304	56.3429
PSDBT	1	8.2579	19.2216	26.6442	32.3998	46.8435
	2	7.4561	17.2292	24.0602	28.9501	41.7997
	10	6.3743	14.2726	19.1060	23.6151	33.7190
	0	10.4555	24.1328	31.8774	40.5039	58.3709
	0.1	10.1294	23.4389	31.1231	39.3894	56.8116
ESDBT	1	8.2760	19.3027	26.6454	32.6058	47.2228
	2	7.4687	17.2925	24.0592	29.1224	42.1307
	10	6.3860	14.3321	19.1083	23.7740	34.0206
	0	10.4291	24.0175	31.8774	40.2191	57.8553
	0.1	10.1048	23.3312	31.1231	39.1219	56.3270
HSDBT	1	8.2575	19.2193	26.6442	32.3935	46.8314
	2	7.4559	17.2278	24.0603	28.9453	41.7898
	10	6.3743	14.2716	19.1060	23.6112	33.7109
	0	10.4242	23.9551	31.8168	40.1759	57.6631
	0.1	10.0997	23.2695	31.0641	39.0759	56.1312
TSDBT	1	8.2489	19.1684	26.5987	32.3545	46.6708
	2	7.4435	17.1739	24.0199	28.9018	41.6412
	10	6.3640	14.2215	19.0726	23.5775	33.6025
	0	10.4555	24.1328	31.8774	40.5039	58.3709
	0.1	10.1294	23.4389	31.1231	39.3894	56.8116
ASDBT	1	8.2760	19.3027	26.6454	32.6058	47.2228
	2	7.4687	17.2925	24.0592	29.1224	42.1307
	10	6.3860	14.3321	19.1083	23.7740	34.0206

Source: Reprinted from International Journal of Mechanical Sciences, Vol 82, Pradhan, K.K., Chakraverty, S., Effects of different shear deformation theories on free vibration of functionally graded beams, Pages 149–160, Copyright (2014), with permission from Elsevier.

Table 4.25 First five non-dimensional frequencies of C-C beams with $L/h = 20$ and different power-law indices (k)

Theory	k	λ_1	λ_2	λ_3	λ_4	λ_5
	0	13.0137	35.7476	69.7005	114.3730	127.5097
	0.1	12.5379	34.4413	67.1564	110.2026	124.4998
CBT	1	10.0377	27.5608	53.7258	88.0841	106.9229
	2	9.1683	25.1637	49.0277	80.3100	96.9243
	10	8.4590	23.2184	45.2357	74.0505	76.8919
	0	12.7979	34.5081	65.7881	105.2642	127.3707
	0.1	12.3352	33.2767	63.4770	101.6280	124.3629
TBT	1	9.8926	26.7404	51.1299	82.0581	106.7716
	2	9.0322	24.4011	46.6244	74.7656	96.7596
	10	8.2939	22.2877	42.3223	67.4185	76.7162
	0	12.7999	34.5229	65.8427	105.4007	127.5097
	0.1	12.3387	33.2992	63.5541	101.8130	124.4983
PSDBT	1	9.8979	26.7600	51.1847	82.1766	106.8755
	2	9.0312	24.3816	46.5538	74.6002	96.8418
	10	8.2673	22.1381	41.8737	66.4469	76.7877
	0	12.8013	34.5310	65.8688	105.4615	127.5097
	0.1	12.3400	33.3067	63.5781	101.8690	124.4983
ESDBT	1	9.8988	26.7653	51.2021	82.2174	106.8758
	2	9.0314	24.3832	46.5601	74.6163	96.8416
	10	8.2671	22.1372	41.8728	66.4481	76.7882
	0	12.7999	34.5229	65.8426	105.4003	127.5097
	0.1	12.3387	33.2992	63.5539	101.8126	124.4983
HSDBT	1	9.8979	26.7600	51.1846	82.1763	106.8755
	2	9.0312	24.3817	46.5543	74.6012	96.8418
	10	8.2674	22.1385	41.8749	66.4493	76.7877
	0	12.7973	34.4991	65.7713	105.1902	127.2672
	0.1	12.3360	33.2763	63.4855	101.6139	124.2622
TSDBT	1	9.8876	26.7252	51.0991	81.9801	106.6921
	2	9.0175	24.3373	46.4513	74.3729	96.6858
	10	8.2586	22.0995	41.7792	66.2071	76.6510
	0	12.8013	34.5310	65.8688	105.4615	127.5097
	0.1	12.3400	33.3067	63.5781	101.8690	124.4983
ASDBT	1	9.8988	26.7653	51.2021	82.2174	106.8758
	2	9.0314	24.3832	46.5601	74.6163	96.8416
	10	8.2671	22.1372	41.8728	66.4481	76.7882

Source: Reprinted from International Journal of Mechanical Sciences, Vol 82, Pradhan, K.K., Chakraverty, S., Effects of different shear deformation theories on free vibration of functionally graded beams, Pages 149–160, Copyright (2014), with permission from Elsevier.

Table 4.26 First five non-dimensional frequencies of S-S beams $L/h = 5$ and different power-law indices (k)

Theory	k	λ_1	λ_2	λ_3	λ_4	λ_5
CBT	0	5.6558	21.6143	31.6911	45.4414	63.3822
	0.1	5.4593	20.8103	30.9565	43.7893	61.9051
	1	4.6636	16.0568	26.7827	34.2260	53.2876
	2	4.4213	14.2766	24.4096	30.6253	48.3421
	10	3.8763	13.2913	19.6543	28.1193	39.1900
TBT	0	5.3817	18.5438	31.6911	35.2056	53.1934
	0.1	5.2002	17.9252	30.9247	34.1177	51.6402
	1	4.4446	14.3143	25.9506	27.9299	42.8712
	2	4.1959	12.8504	23.2057	25.2823	38.9005
	10	3.6384	11.5197	18.7881	21.9168	32.9732
PSDBT	0	5.3060	18.4831	31.0642	35.4865	54.0823
	0.1	5.1273	17.8706	30.3157	34.3947	52.5302
	1	4.3813	14.2174	25.5089	28.0437	43.4188
	2	4.1278	12.6923	22.8017	25.1730	39.0039
	10	3.5577	11.2350	18.3975	21.3698	32.1831
ESDBT	0	5.3949	18.6795	31.6911	35.6820	54.2074
	0.1	5.2142	18.0668	30.9259	34.6085	52.6748
	1	4.4552	14.3970	25.9777	28.2637	43.6356
	2	4.1948	12.8510	23.1896	25.3619	39.1906
	10	3.6104	11.3330	18.7257	21.4274	32.1755
HSDBT	0	5.3908	18.6339	31.6911	35.5312	53.8668
	0.1	5.2104	18.0247	30.9255	34.4686	52.3575
	1	4.4519	14.3692	25.9679	28.1592	43.3837
	2	4.1921	12.8307	23.1794	25.2751	38.9665
	10	3.6078	11.3103	18.7187	21.3390	31.9603
TSDBT	0	5.3920	18.6507	31.6911	35.5825	54.0031
	0.1	5.2115	18.0400	30.9257	34.5152	52.4833
	1	4.4529	14.3795	25.9715	28.1948	43.4848
	2	4.1927	12.8376	23.1824	25.3039	39.0557
	10	3.6084	11.3174	18.7209	21.3674	32.0435
ASDBT	0	5.3949	18.6795	31.6911	35.6820	54.2074
	0.1	5.2142	18.0668	30.9259	34.6085	52.6748
	1	4.4552	14.3970	25.9777	28.2637	43.6356
	2	4.1948	12.8510	23.1896	25.3619	39.1906
	10	3.6104	11.3330	18.7257	21.4274	32.1755

Source: Reprinted from International Journal of Mechanical Sciences, Vol 82, Pradhan, K.K., Chakraverty, S., Effects of different shear deformation theories on free vibration of functionally graded beams, Pages 149–160, Copyright (2014), with permission from Elsevier.

Table 4.27 First five non-dimensional frequencies of S-S beams with $L/h = 20$ and different power-law indices (k)

Theory	k	λ_1	λ_2	λ_3	λ_4	λ_5
CBT	0	5.7422	22.8985	51.2610	90.4935	126.7643
	0.1	5.5420	22.0597	49.3956	87.1663	123.7531
	1	4.7341	17.5982	39.7292	68.9932	105.2888
	2	4.4900	16.0377	36.3937	62.4870	94.9183
	10	3.9382	14.8299	33.4081	57.8516	76.3703
TBT	0	5.7225	22.5921	49.7814	86.1098	126.7643
	0.1	5.5234	21.7722	48.0054	83.0516	123.7364
	1	4.7184	17.4005	38.7474	66.2688	101.2400
	2	4.4737	15.8563	35.4787	60.0715	92.1499
	10	3.9207	14.6025	32.2938	54.8651	75.8946
PSDBT	0	5.7227	22.5947	49.7970	86.1498	126.7643
	0.1	5.5237	21.7768	48.0299	83.1175	123.7367
	1	4.7186	17.4023	38.7579	66.2939	101.2959
	2	4.4729	15.8473	35.4368	59.9594	91.8916
	10	3.9175	14.5613	32.1025	54.3573	75.8240
ESDBT	0	5.7228	22.5969	49.8072	86.1805	126.7643
	0.1	5.5238	21.7787	48.0392	83.1456	123.7368
	1	4.7187	17.4037	38.7647	66.3132	101.3427
	2	4.4729	15.8477	35.4393	59.9666	91.9121
	10	3.9175	14.5611	32.1019	54.3574	75.8245
HSDBT	0	5.7227	22.5947	49.7970	86.1496	126.7643
	0.1	5.5237	21.7768	48.0298	83.1173	123.7367
	1	4.7186	17.4023	38.7579	66.2938	101.2955
	2	4.4729	15.8473	35.4370	59.9598	91.8925
	10	3.9175	14.5614	32.1029	54.3585	75.8242
TSDBT	0	5.7225	22.5922	49.7846	86.1171	126.7643
	0.1	5.5235	21.7744	48.0181	83.0864	123.7366
	1	4.7184	17.4007	38.7495	66.2739	101.2624
	2	4.4727	15.8453	35.4265	59.9362	91.8476
	10	3.9172	14.5582	32.0867	54.3220	75.8189
ASDBT	0	5.7228	22.5969	49.8072	86.1805	126.7643
	0.1	5.5238	21.7787	48.0392	83.1456	123.7368
	1	4.7187	17.4037	38.7647	66.3132	101.3427
	2	4.4729	15.8477	35.4393	59.9666	91.9121
	10	3.9175	14.5611	32.1019	54.3574	75.8245

Source: Reprinted from International Journal of Mechanical Sciences, Vol 82, Pradhan, K.K., Chakraverty, S., Effects of different shear deformation theories on free vibration of functionally graded beams, Pages 149–160, Copyright (2014), with permission from Elsevier.

Table 4.28 First five non-dimensional frequencies of C-F beams with $L/h = 5$ and different power-law indices (k)

Theory	k	λ_1	λ_2	λ_3	λ_4	λ_5
CBT	0	2.0321	12.1896	15.9398	32.0486	47.8188
	0.1	1.9578	11.7470	15.5652	30.8969	46.6963
	1	1.5664	9.3464	13.4228	24.4870	40.3255
	2	1.4297	8.4787	12.2129	22.1053	36.7305
	10	1.3192	7.8076	9.6956	20.3742	29.1266
TBT	0	1.9827	10.6377	15.9077	25.2861	41.9246
	0.1	1.9115	10.2834	15.5329	24.4931	40.6758
	1	1.5332	8.3236	13.3652	20.0055	33.4633
	2	1.3986	7.5521	12.1359	18.1154	30.2444
	10	1.2817	6.7227	9.6211	15.8025	25.9437
PSDBT	0	1.9486	10.5856	15.6245	25.4327	42.5695
	0.1	1.8788	10.2376	15.2564	24.6530	41.3403
	1	1.5071	8.2692	13.1282	20.0671	33.8649
	2	1.3738	7.4642	11.9190	18.0228	30.2913
	10	1.2547	6.5546	9.4480	15.4059	25.3167
ESDBT	0	1.9838	10.6886	15.9268	25.5342	42.5491
	0.1	1.9131	10.3501	15.5642	24.8010	41.4340
	1	1.5350	8.3688	13.3909	20.2180	34.0074
	2	1.3989	7.5484	12.1549	18.1439	30.3994
	10	1.2764	6.6047	9.6353	15.4431	25.2875
HSDBT	0	1.9836	10.6852	15.9398	25.5118	42.4810
	0.1	1.9127	10.3367	15.5642	24.7379	41.2697
	1	1.5348	8.3590	13.3906	20.1702	33.8804
	2	1.3988	7.5424	12.1549	18.1095	30.2991
	10	1.2763	6.5998	9.6349	15.4114	25.1959
TSDBT	0	1.9829	10.6460	15.8816	25.3436	42.1192
	0.1	1.9119	10.2995	15.5075	24.5767	40.9193
	1	1.5325	8.3267	13.3453	20.0423	33.5970
	2	1.3958	7.5084	12.1148	17.9843	30.0304
	10	1.2743	6.5650	9.6006	15.2868	24.9558
ASDBT	0	1.9835	10.6739	15.9077	25.4682	42.4061
	0.1	1.9126	10.3256	15.5329	24.6940	41.1909
	1	1.5337	8.3480	13.3657	20.1335	33.8121
	2	1.3973	7.5277	12.1327	18.0649	30.2204
	10	1.2753	6.5842	9.6162	15.3660	25.1292

Source: Reprinted from International Journal of Mechanical Sciences, Vol 82, Pradhan, K.K., Chakraverty, S., Effects of different shear deformation theories on free vibration of functionally graded beams, Pages 149–160, Copyright (2014), with permission from Elsevier.

Table 4.29 First five non-dimensional frequencies of C-F beams with $L/h = 20$ and different power-law indices (k)

Theory	k	λ_1	λ_2	λ_3	λ_4	λ_5
CBT	0	2.0468	12.7899	35.6470	63.7593	69.3888
	0.1	1.9719	12.3224	34.3452	62.2520	66.8600
	1	1.5785	9.8618	27.4776	52.9441	53.9887
	2	1.4419	9.0053	25.0796	47.9763	49.2651
	10	1.3304	8.3091	23.1409	38.3908	45.0313
TBT	0	2.0435	12.6516	34.7679	63.6307	66.3961
	0.1	1.5753	9.7654	26.8859	51.3721	53.4646
	1	1.5753	9.7654	26.8859	51.3721	53.4646
	2	1.4384	8.9129	24.5248	46.7401	48.5580
	10	1.3272	8.2010	22.4728	38.3100	42.7970
PSDBT	0	2.0435	12.6546	34.7893	63.7593	66.4727
	0.1	1.9689	12.1964	33.5456	62.2492	64.1410
	1	1.5764	9.7738	26.9179	51.4314	53.5848
	2	1.4397	8.9189	24.5329	46.7072	48.6656
	10	1.3274	8.1880	22.3813	38.3687	42.4996
ESDBT	0	2.0435	12.6554	34.7947	63.7593	66.4912
	0.1	1.9689	12.1972	33.5506	62.2493	64.1579
	1	1.5764	9.7743	26.9214	51.4429	53.5857
	2	1.4397	8.9190	24.5338	46.7105	48.6660
	10	1.3274	8.1878	22.3802	38.3687	42.4972
HSDBT	0	2.0435	12.6546	34.7893	63.7593	66.4727
	0.1	1.9689	12.1964	33.5456	62.2492	64.1409
	1	1.5764	9.7738	26.9179	51.4314	53.5848
	2	1.4397	8.9189	24.5330	46.7076	48.6656
	10	1.3274	8.1880	22.3816	38.3687	42.5006
TSDBT	0	2.0434	12.6496	34.7538	63.5262	66.3470
	0.1	1.9688	12.1914	33.5114	62.0245	64.0194
	1	1.5745	9.7589	26.8624	51.3310	53.3706
	2	1.4371	8.8993	24.4639	46.6040	48.4560
	10	1.3261	8.1751	22.3249	38.2562	42.3271
ASDBT	0	2.0435	12.6554	34.7947	63.7593	66.4912
	0.1	1.9689	12.1972	33.5506	62.2493	64.1579
	1	1.5764	9.7743	26.9214	51.4429	53.5857
	2	1.4397	8.9190	24.5338	46.7105	48.6660
	10	1.3274	8.1878	22.3802	38.3687	42.4972

Source: Reprinted from International Journal of Mechanical Sciences, Vol 82, Pradhan, K.K., Chakraverty, S., Effects of different shear deformation theories on free vibration of functionally graded beams, Pages 149–160, Copyright (2014), with permission from Elsevier.

reported here, one may easily find the non-dimensional frequencies of FG beam taking any slenderness ratio and power-law index subjected to any set of edge supports using the above-mentioned SDBTs.

Instead of assuming all SDBTs of Eq. (2.5), ascending values of the power-law index (n) in PESDBT are considered to report the first five non-dimensional frequencies for two different slenderness ratios ($L/h = 5, 20$) and a fixed power-law index ($k = 1$) in Tables 4.30 and 4.31 with all sets of classical edge supports. The material properties of FG constituents are: $E_m = 70$ GPa, $\rho_m = 2702\,\text{kg/m}^3$, $E_c = 380$ GPa, $\rho_c = 3960\,\text{kg/m}^3$, and $\nu_m = \nu_c = 0.3$. In Tables 4.30 and 4.31, % Difference $= \left|\frac{\lambda_{CBT}-\lambda_{n=200}}{\lambda_{CBT}}\right| \times 100$ is meant for the percentage errors generated by comparing the results computed for CBT and PESDBT with $n = 200$. It can easily be noticed here that frequencies obtained in the case of PESDBT with $n = 0$ will follow the behavior of the **Euler-Bernoulli FG beam.**[5] Looking into the ascending behavior of the results, a little fluctuation in non-dimensional frequencies at each mode can be observed while considering the very large value of n. It is also evident that frequencies will gradually coincide with the Euler-Bernoulli beam for a very large value of n.

Analyzing Tables 4.24–4.29 for the first five non-dimensional frequencies of FG beams with different sets of edge supports, one may plot the effects of power-law indices (k) on non-dimensional frequencies for a given slenderness ratio (L/h) within the framework of all SDBTs of Eq. (2.4). But the present study includes the plot of effects of power-law indices on fundamental and fifth non-dimensional frequencies of C-C FG beams respectively in Figs. 4.1 and 4.2 with $L/h = 5$ and 20. It can easily be observed that the non-dimensional frequencies are gradually decreasing with an increase in power-law indices and are increasing with an increase in slenderness ratios. For $L/h = 5$, the difference among the frequencies at each mode is quite significant while comparing the results obtained from these mentioned SDBTs, whereas these are found to be negligible for $L/h = 20$, irrespective of the beam theories considered.

4.2 CONCLUDING REMARKS

In this investigation, we have studied vibration characteristics of FG beams subjected to various sets of boundary conditions within the framework of different SDBTs. This study may be the first of its kind to the best of

[5] Analysis of FG beam based on classical or Euler-Bernoulli beam theory.

Table 4.30 Effect of power-law exponents (n) on frequencies of functionally graded beams with $L/h = 5$ and $k = 1$

BCs	Mode	Power-law exponent (n) in PESDBT 1	3	5	10	30	50	100	200	CBT	% Difference
	1	7.9680	7.9968	8.0314	8.0735	8.1126	8.1218	8.1290	8.1327	9.3755	13.2558
	2	18.6207	18.6806	18.7983	18.9487	19.0920	19.1259	19.1525	19.1662	24.2623	21.0042
C-C	3	25.3862	25.3919	25.3962	25.4012	25.4058	25.4068	25.4076	25.4080	25.6670	1.0091
	4	31.5143	31.5724	31.8034	32.1079	32.4022	32.4719	32.5269	32.5553	44.2130	26.3671
	5	45.5201	45.4909	45.8407	46.3218	46.7952	46.9081	46.9971	47.0430	51.2429	8.1961
	1	5.9349	5.9485	5.9646	5.9842	6.0024	6.0066	6.0100	6.0117	6.5431	8.1215
	2	16.3317	16.3738	16.4478	16.5423	16.6324	16.6536	16.6702	16.6788	19.5487	14.6808
C-S	3	24.9772	24.9903	25.0047	25.0221	25.0381	25.0419	25.0448	25.0463	25.6086	2.1957
	4	29.6084	29.6648	29.8403	30.0710	30.2942	30.3470	30.3886	30.4100	38.2848	20.5689
	5	43.8186	43.8445	44.1353	44.5299	44.9159	45.0075	45.0795	45.1166	50.9684	11.4812
	1	1.4645	1.4655	1.4664	1.4674	1.4683	1.4685	1.4686	1.4687	1.4925	1.5946
	2	8.0125	8.0354	8.0600	8.0887	8.1151	8.1212	8.1262	8.1284	8.9080	8.7517
C-F	3	12.7328	12.7337	12.7345	12.7354	12.7362	12.7364	12.7365	12.7366	12.7591	0.1763
	4	19.3830	19.4556	19.5516	19.6662	19.7729	19.7974	19.8178	19.8267	23.3377	15.0443
	5	32.6162	32.7413	32.9500	33.2040	33.4429	33.4978	33.5441	33.5638	38.3349	12.4458

Continued

Table 4.30 Effect of power-law exponents (n) on frequencies of functionally graded beams with $L/h = 5$ and $k = 1$—cont'd

BCs	Mode	Power-law exponent (n) in PESDBT									
		1	3	5	10	30	50	100	200	CBT	% Difference
S–S	1	4.2632	4.2663	4.2712	4.2775	4.2834	4.2848	4.2858	4.2864	4.4488	3.6504
	2	13.8327	13.8534	13.8902	13.9379	13.9835	13.9942	14.0027	14.0070	15.3172	8.5538
	3	24.8238	24.8358	24.8519	24.8722	24.8916	24.8961	24.8997	24.9016	25.5491	2.5343
	4	27.2748	27.3240	27.4470	27.6093	27.7662	27.8032	27.8323	27.8473	32.6495	14.7083
	5	42.2306	42.2733	42.5149	42.8446	43.1690	43.2460	43.3065	43.3378	50.8326	14.7441
S–F	1	5.9733	5.9786	5.9849	5.9928	6.0001	6.0018	6.0031	6.0038	6.1998	3.1614
	2	12.5037	12.5070	12.5100	12.5135	12.5167	12.5175	12.5180	12.5183	12.6055	0.6918
	3	17.1015	17.1447	17.1973	17.262	17.3226	17.3367	17.3479	17.3536	19.1106	9.1939
	4	30.5977	30.7021	30.8467	31.0264	31.1954	31.2349	31.2661	31.2820	35.9094	12.8863
	5	37.2466	37.2783	37.3066	37.3394	37.3693	37.3763	37.3818	37.3846	38.6742	3.3345
F–F	1	8.4865	8.4993	8.5095	8.5209	8.5314	8.5338	8.5356	8.5366	8.8114	3.1187
	2	20.0769	20.1558	20.2223	20.2987	20.368	20.3841	20.3966	20.4031	22.3449	8.6901
	3	25.3277	25.3353	25.3412	25.3478	25.3537	25.3550	25.3561	25.3566	25.4925	0.5331
	4	33.9054	34.0842	34.2529	34.4504	34.6309	34.6728	34.7058	34.2275	39.9919	14.4139
	5	47.3796	47.6262	47.8741	48.1557	48.4017	48.4571	48.5001	48.5221	51.1175	5.0773

Source: Reprinted from Applied Mathematics and Computation, Vol 268, Pradhan, K.K., Chakraverty, S., Generalized power-law exponent based shear deformation theory for free vibration of functionally graded beams, Pages 1240–1258, Copyright (2015), with permission from Elsevier.

Table 4.31 Effect of power-law exponents (n) on frequencies of functionally graded beams with $L/h = 20$ and $k = 1$

BCs	Mode	Power-law exponent (n) in PESDBT								CBT	% Difference
		1	3	5	10	30	50	100	200		
C–C	1	9.4438	9.4490	9.4529	9.4574	9.4614	9.4623	9.4630	9.4634	9.5667	1.0798
	2	25.5599	25.5890	25.6115	25.6369	25.6597	25.6649	25.6691	25.6712	26.2740	2.2943
	3	48.9471	49.0359	49.1051	49.1837	49.2544	49.2707	49.2835	49.2901	51.2050	3.7397
	4	78.6623	78.8616	79.0201	79.2007	79.3636	79.4013	79.4308	79.4460	83.9755	5.3938
	5	101.781	101.782	101.783	101.784	101.785	101.785	101.786	101.786	101.811	0.0246
C–S	1	6.6254	6.6274	6.6290	6.6308	6.6324	6.6327	6.6330	6.6331	6.6738	0.6098
	2	20.9569	20.9727	20.9850	20.9989	21.0114	21.0143	21.0166	21.0177	21.3447	1.5319
	3	42.7407	42.7980	42.8429	42.8939	42.9397	42.9503	42.9586	42.9629	44.1849	2.7657
	4	71.0299	71.1701	71.2820	71.4095	71.5244	71.5510	71.5718	71.5825	74.7113	4.1879
	5	100.002	100.049	100.083	100.121	100.153	100.160	100.166	100.169	100.729	0.5559
C–F	1	1.5021	1.5022	1.5023	1.5024	1.5024	1.5024	1.5024	1.5024	1.5040	0.1064
	2	9.317	9.3204	9.3230	9.3259	9.3284	9.3290	9.3295	9.3297	9.3964	0.7098
	3	25.6725	25.6938	25.7099	25.7282	25.7445	25.7483	25.7512	25.7528	26.1802	1.6325
	4	49.1173	49.1844	49.2354	49.2930	49.3445	49.3562	49.3656	49.3704	50.4900	2.2175
	5	50.9201	50.9240	50.9272	50.9310	50.9347	50.9356	50.9362	50.9367	51.2900	0.6888

Continued

Table 4.31 Effect of power-law exponents (n) on frequencies of functionally graded beams with $L/h = 20$ and $k = 1$—cont'd

BCs	Mode	Power-law exponent (n) in PESDBT								CBT	% Difference
		1	3	5	10	30	50	100	200		
S-S	1	4.5024	4.5030	4.5034	4.5039	4.5044	4.5045	4.5045	4.5046	4.5161	0.2546
	2	16.6154	16.6224	16.6279	16.6341	16.6397	16.641	16.6420	16.6425	16.7876	0.8643
	3	37.0436	37.0773	37.1042	37.1348	37.1623	37.1686	37.1736	37.1762	37.8993	1.9079
	4	63.4363	63.5279	63.6015	63.6854	63.7609	63.7784	63.7920	63.7991	65.8153	3.0634
	5	97.0898	97.2979	97.4674	97.6611	97.8359	97.8763	97.9080	97.9243	100.439	2.5112
S-F	1	6.5489	6.5498	6.5504	6.5512	6.5519	6.5520	6.5522	6.5522	6.5691	0.2573
	2	20.9383	20.9481	20.9556	20.9641	20.9717	20.9734	20.9748	20.9755	21.1721	0.9286
	3	42.3909	42.4285	42.4574	42.4900	42.5193	42.5261	42.5314	42.5341	43.2930	1.7529
	4	50.7788	50.7853	50.7904	50.7962	50.8014	50.8026	50.8004	50.8041	50.9527	0.2916
	5	71.6894	71.8032	71.8921	71.9929	72.0835	72.1045	72.1209	72.1293	74.5611	3.2615
F-F	1	9.4849	9.4862	9.4873	9.4884	9.4894	9.4897	9.4898	9.4899	9.5160	0.2743
	2	25.7549	25.7683	25.7783	25.7895	25.7996	25.8019	25.8037	25.8046	26.0640	0.9952
	3	49.4271	49.4791	49.5181	49.5619	49.6012	49.6103	49.6174	49.6210	50.6546	2.0405
	4	79.6003	79.7349	79.8368	79.9519	80.0553	80.0792	80.0979	80.1075	82.8799	3.3451
	5	101.411	101.414	101.416	101.419	101.421	101.421	101.422	101.422	101.469	0.0463

Source: Reprinted from Applied Mathematics and Computation, Vol 268, Pradhan, K.K., Chakraverty, S., Generalized power-law exponent based shear deformation theory for free vibration of functionally graded beams, Pages 1240–1258, Copyright (2015), with permission from Elsevier.

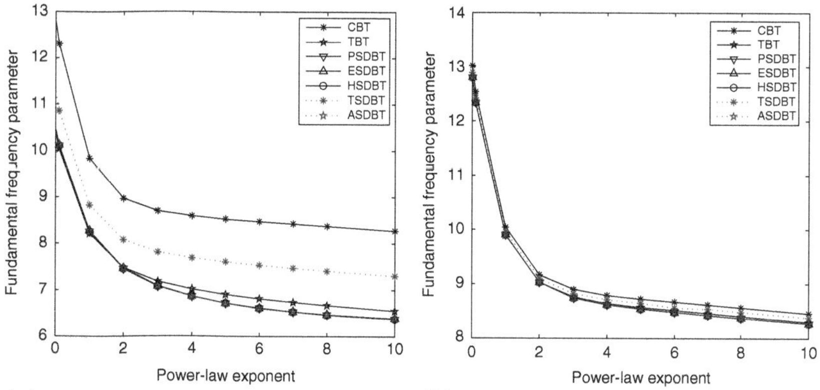

(a) $L/h = 5$ (b) $L/h = 20$

Figure 4.1 Effect of power-law indices on fundamental non-dimensional frequencies of C-C FG beams using all SDBTs: (a) $L/h = 5$; (b) $L/h = 20$. (*Reprinted from International Journal of Mechanical Sciences, Vol 82, Pradhan, K.K., Chakraverty, S., Effects of different shear deformation theories on free vibration of functionally graded beams, Pages 149–160, Copyright (2014), with permission from Elsevier.*).

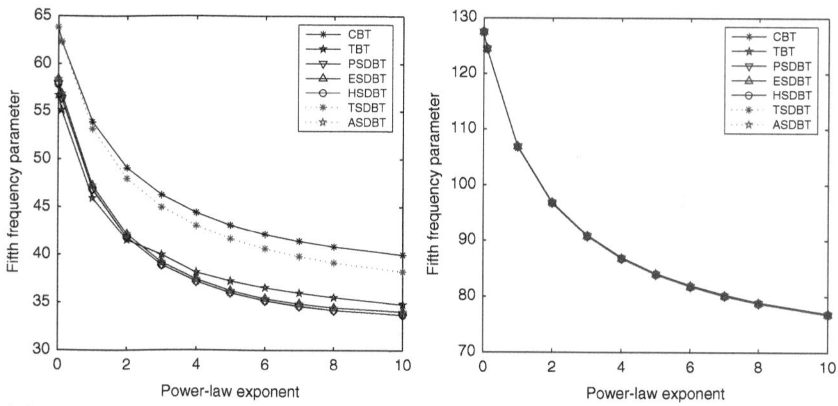

(a) $L/h = 5$ (b) $L/h = 20$

Figure 4.2 Effect of power-law indices on fifth non-dimensional frequencies of C-C FG beams using all SDBTs: (a) $L/h = 5$; (b) $L/h = 20$. (*Reprinted from International Journal of Mechanical Sciences, Vol 82, Pradhan, K.K., Chakraverty, S., Effects of different shear deformation theories on free vibration of functionally graded beams, Pages 149–160, Copyright (2014), with permission from Elsevier.*).

the authors' knowledge about the Rayleigh-Ritz method in FG beams to handle any sets of boundary conditions. But one has to keep track of the convergence pattern to report the corresponding results. By analyzing the above formulations and numerical results, the following conclusions may be drawn.

- In the Rayleigh-Ritz method, increase in the number of polynomials (n) is a key factor in deciding the convergence of the non-dimensional frequencies.
- The slenderness ratios, power-law variation of material properties, different material distributions, and different SDBTs play crucial roles when examining the vibration characteristics of FG beams.
- The non-dimensional frequencies are increasing with an increase in L/h ratios and decreasing with an increase in k regardless of the beam theories considered.
- While comparing results based on various SDBTs, the difference among the non-dimensional frequencies at each mode is significant for $L/h <$ 20, whereas it becomes gradually negligible for $L/h \geq 20$.
- Other SDBTs can also be extended easily in the above analysis.

CHAPTER 5

Vibration Problems of Functionally Graded Rectangular Plates

The vibration characteristics of isotropic as well as functionally graded (FG) rectangular plates with specific (or arbitrary) geometrical configurations have been exhaustively carried out in earlier decades, but such studies on thin FG plates are scarce. Major findings on numerical solutions of such problems are given below.

Free vibration frequencies of rectangular plates subject to different possible combinations of classical boundary conditions have been evaluated in Leissa (1973), Bhat (1985), and Singh and Chakraverty (1994a) by means of the Rayleigh-Ritz method. Roque et al. (2007) have examined free vibration of FG plates based on a refined theory by the multiquadric radial basis function method. Reddy (2000) has presented a general formulation for FG plates using third-order shear deformation plate theory. Free vibration of moderately thick rectangular plates has been investigated by Liu and Liew (1999) using the differential quadrature element method. Yang and Shen (2001) have proposed the dynamic response of initially stressed FG rectangular thin plates subject to distributed impulsive lateral loads resting on an elastic foundation. Natural frequencies of square FG plates have been computed by Ferreira et al. (2006) using the asymmetric collocation method with multiquadric basis functions. It is clearly mentioned in Abrate (2008) that FG plates behave like homogeneous plates using the higher-order plate theory. Matsunaga (2008) has estimated the natural frequencies and buckling stresses of FG plates based on a 2D higher-order shear deformation theory. Exact closed-form solutions of 3D elasticity theory are presented in Hosseini-Hashemi et al. (2013) to study both in-plane and out-of-plane free vibrations for FG simply supported rectangular plates.

In view of the above, free vibration of thin FG rectangular plates subjected to various possible boundary conditions based on classical plate theory is investigated in this chapter. Material properties of the FG plate constituents vary continuously along thickness direction in either

Vibration of Functionally Graded Beams and Plates
http://dx.doi.org/10.1016/B978-0-12-804228-1.00005-5

power-law or exponential law. Trial functions denoting the displacement components are expressed as linear combination of simple algebraic polynomials, which may handle any combination of boundary conditions easily. The generalized eigenvalue problem can be obtained by means of the Rayleigh-Ritz method (as stated in Chapter 3). An excellent validation of natural frequencies can be observed after the test of convergence with respect to number of polynomials. The effects of constituent volume fractions, aspect ratios, and power-law indices on the natural frequencies can also be found here. Three-dimensional (3D) mode shapes are depicted for six lowest modes of different FG rectangular plates.

5.1 NUMERICAL MODELING

In this section, the admissible function of FG rectangular plate is to be assumed in the Rayleigh-Ritz method rather than that of other geometries. The numerical formulation for free vibration of FG thin rectangular plates holds the step-by-step Rayleigh-Ritz procedures mentioned in Section 3.3.2 to find the generalized eigenvalue problem of the form Eq. (3.26). The next section mainly deals with numerical results obtained for the above said problem with both the variation patterns of FG material properties. As such, test of convergence and comparison of natural frequencies with available results in special cases have been carried out to report the new results. Consequently, 3D mode shapes of FG plates with a few edge supports have also been demonstrated. The detail of these computation and demonstration for power-law and exponential gradation forms have already mentioned in Chakraverty and Pradhan (2014b,a).

5.2 CONVERGENCE AND COMPARISON STUDIES

The convergence of non-dimensional frequencies of FG rectangular plates has been checked with an increase in the number of polynomials involved in the displacement components. The material properties for this investigation are considered as shown in Table 5.1.

The following expression is meant for computing the non-dimensional frequencies as

$$\lambda = \omega a^2 \sqrt{\frac{\rho_c h}{D_c}} \tag{5.1}$$

Table 5.1 Material properties of the FG plate constituents

Properties	Unit	Aluminum	Al_2O_3	SUS304	Si_3N_4
E	GPa	70	380	208	322
ρ	kg/m^3	2700	3800	8166	2370
ν	-	0.3	0.3	0.3	0.3

Source: Reproduced from Chakraverty, S., Pradhan, K.K., Free vibration of exponential functionally graded rectangular plates in thermal environment with general boundary conditions. Aerospace Science and Technology 2014;36:132–156. Copyright © 2015 Elsevier Masson SAS. All rights reserved.

where $D_{\c{c}}$ (flexural rigidity) $= \frac{E_c h^3}{12(1-\nu^2)}$ of the FG plate due to the deformation effect.

In the case of power law gradation, convergence of the first seven non-dimensional frequencies of isotropic Al/Al_2O_3 square plate (assuming power-law index as 0 and aspect ratio as unity) with different edge conditions, viz. CCCC, CSCS, and SSSS, is performed in Table 5.2 with an increase in the number of polynomials involved in displacement components. Clamped isotropic $SUS304/Si_3N_4$ rectangular plate with various edge ratio (a/b), viz. 2/5, 2/3, 1.0, 1.5, are considered in Table 5.3 to study the convergence of frequencies. It may be observed in Tables 5.2 and 5.3 that non-dimensional frequencies for isotropic plates converge gradually with an increase in the number of polynomials.

On the other hand, the exponential gradation criterion in FG square Al/Al_2O_3 plate with different edge conditions, viz. CCCC, CSCS, and SSSS, are considered in Table 5.4 and clamped FG rectangular $SUS304/Si_3N_4$ plate with various edge ratio (a/b), viz. 2/5, 2/3, 1.0, 1.5, are assumed in Table 5.5 to analyze the convergence of natural frequencies. It may also be observed in Tables 5.4 and 5.5 that non-dimensional frequencies for isotropic plates converge gradually with an increase in the number of polynomials in the case of exponential variation of material properties. It may also be concluded that the number of polynomials involved in displacement components play a major role in the convergence of non-dimensional frequencies irrespective of the edge condition, gradation criteria, and configuration of the plate considered.

After checking the convergence, a comparison of non-dimensional frequencies with the available literature may easily be done. In this regard, the validation is carried out with existing ones for isotropic cases of FG rectangular (or square) plates, which considers $k = 0$ in power-law variation and $\delta = 0$ in exponential gradation. In Table 5.6, a comparison of the

Table 5.2 Convergence of non-dimensional frequencies for isotropic square Al/Al_2O_3 plate with $k = 0$ in power-law variation

BCs	Sources	λ_1	λ_2	λ_3	λ_4	λ_5	λ_6	λ_7
	10 × 10	36.0000	73.4329	73.4329	108.5910	137.2938	138.6506	168.8140
	13 × 13	35.9894	73.4329	73.4329	108.4270	132.1428	137.6999	168.8140
CCCC	16 × 16	35.9888	73.4172	73.4329	108.2653	131.8982	132.4236	168.8139
	19 × 19	35.9888	73.4039	73.4152	108.2653	131.8982	132.4236	165.2066
	20 × 20	35.9888	73.3989	73.4152	108.2653	131.8982	132.4236	165.2066
	10 × 10	28.9525	54.8846	69.3469	100.8762	135.1902	142.5894	163.3314
	13 × 13	28.9514	54.8846	69.3469	94.9189	103.7179	135.0244	163.3314
CSCS	16 × 16	28.9509	54.7571	69.3469	94.7034	103.7178	129.3120	163.2495
	19 × 19	28.9509	54.7484	69.3439	94.7034	103.7178	129.3120	141.3811
	20 × 20	28.9509	54.7439	69.3439	94.7034	103.7178	129.3120	141.3811
	10 × 10	19.7449	49.5132	49.5132	92.5635	139.5994	139.9505	168.7293
	13 × 13	19.7408	49.5132	49.5132	86.0902	100.1799	139.5995	168.7293
SSSS	16 × 16	19.7392	49.3736	49.5132	79.4007	100.1729	100.1868	168.6555
	19 × 19	19.7392	49.3498	49.4909	79.4007	100.1729	100.1868	130.3895
	20 × 20	19.7392	49.3490	49.4909	79.4007	100.1729	100.1868	130.3895

Source: First published in Chakraverty, S., Pradhan, K.K., 2014. Free vibration of functionally graded thin rectangular plates resting on Winkler elastic foundation with general boundary conditions using Rayleigh-Ritz method. Int. J. Appl. Mech. 06, 1450043.

Table 5.3 Convergence of non-dimensional frequencies for CCCC isotropic rectangular SUS304/Si_3N_4 plate with $k = 0$ in power-law variation

a/b	Sources	λ_1	λ_2	λ_3	λ_4	λ_5	λ_6	λ_7
	7 × 7	23.6482	27.8269	35.8006	63.2107	68.4171	129.0781	223.3135
	9 × 9	23.6482	27.8171	35.8006	63.1294	68.4171	75.5319	129.0781
2/5	11 × 11	23.6478	27.8151	35.8006	48.8897	63.1294	68.4171	75.5319
	13 × 13	23.6467	27.8151	35.7980	48.8897	63.1294	67.4815	75.5319
	14 × 14	23.6467	27.8151	35.7980	48.8897	63.1294	67.4377	75.5319
	7 × 7	27.0132	41.8994	66.2992	68.7006	80.3935	131.8556	225.9695
	9 × 9	27.0132	41.8859	66.1653	68.7006	80.3935	104.1890	131.8556
2/3	11 × 11	27.0128	41.7228	66.1653	68.7006	80.3935	104.1890	109.3627
	13 × 13	27.0082	41.7228	66.1653	68.5959	79.8459	104.1890	109.3627
	14 × 14	27.0082	41.7228	66.1653	68.5959	79.8454	104.1890	109.3627
	7 × 7	36.0000	73.5504	74.2967	108.5910	137.2938	138.6506	231.5822
	9 × 9	36.0000	73.4329	74.1843	108.5910	137.2938	138.6506	168.8140
1.0	11 × 11	35.9995	73.4329	73.4329	108.5910	132.3887	138.0391	168.8140
	13 × 13	35.9894	73.4329	73.4329	108.4270	132.1428	137.6999	168.8140
	14 × 14	35.9894	73.4329	73.4329	108.2653	132.1428	137.6999	168.8140
	7 × 7	60.7797	93.9085	151.4183	154.5763	180.8853	246.0817	296.6751
	9 × 9	60.7797	93.8763	151.1258	154.5763	180.8853	234.4252	246.0662
1.5	11 × 11	60.7782	93.8763	148.8718	150.1498	180.8853	234.4252	246.0662
	13 × 13	60.7678	93.8763	148.8718	149.9028	180.8846	234.4252	246.0662
	14 × 14	60.7678	93.8763	148.8718	149.9028	179.6522	234.4252	246.0662

Source: First published in Chakraverty, S., Pradhan, K.K., 2014. Free vibration of functionally graded thin rectangular plates resting on Winkler elastic foundation with general boundary conditions using Rayleigh-Ritz method. Int. J. Appl. Mech. 06, 1450043.

Table 5.4 Convergence of non-dimensional frequencies for FG Al/Al_2O_3 square plates with exponential gradation

BCs	Sources	λ_1	λ_2	λ_3	λ_4	λ_5	λ_6	λ_7
CCCC	7 × 7	28.3505	57.9220	58.5097	85.5170	108.1209	109.1894	182.3744
	9 × 9	28.3505	57.8295	58.4213	85.5170	108.1209	109.1894	132.9435
	11 × 11	28.3501	57.8295	57.8295	85.5170	104.2581	108.7078	132.9435
	13 × 13	28.3422	57.8295	57.8295	85.3879	104.0644	108.4407	132.9435
	14 × 14	28.3422	57.8295	57.8295	85.2605	104.0644	108.4407	132.9435
CSCS	7 × 7	22.8005	43.3085	55.4625	79.4415	106.4643	112.2913	224.8264
	9 × 9	22.8005	43.2224	55.2722	79.4415	106.4643	112.2913	128.6259
	11 × 11	22.7996	43.2224	54.6117	79.4415	81.8807	106.4717	128.6259
	13 × 13	22.7996	43.2224	54.6117	74.7500	81.6794	106.3337	128.6259
	14 × 14	22.7996	43.2224	54.6117	74.5803	81.6794	106.3337	128.6259
SSSS	7 × 7	15.5494	39.2356	46.4567	72.8951	109.9366	110.2131	223.1847
	9 × 9	15.5494	38.9923	46.2328	72.8951	109.9366	110.2131	132.8768
	11 × 11	15.5482	38.9923	38.9923	72.8951	79.0228	110.0754	132.8768
	13 × 13	15.5462	38.9923	38.9923	67.7973	78.8931	109.9367	132.8768
	14 × 14	15.5462	38.9923	38.9923	62.5293	78.8931	109.9367	132.8768

Source: Reproduced from Chakraverty, S., Pradhan, K.K., Free vibration of exponential functionally graded rectangular plates in thermal environment with general boundary conditions. Aerospace Science and Technology 2014;36:132–156.

Table 5.5 Convergence of non-dimensional frequencies for CCCC FG SUS304/Si_3N_4 rectangular plates with exponential gradation

a/b	Sources	λ_1	λ_2	λ_3	λ_4	λ_5	λ_6	λ_7
	7 × 7	15.1802	17.8626	22.9811	40.5761	43.9182	82.8577	143.3492
	9 × 9	15.1802	17.8563	22.9811	40.5239	43.9182	48.4853	82.8577
2/5	11 × 11	15.1800	17.8550	22.9811	31.3832	40.5239	43.9182	48.4853
	13 × 13	15.1793	17.8550	22.9794	31.3832	40.5239	43.3177	48.4853
	14 × 14	15.1793	17.8550	22.9794	31.3832	40.5239	43.2896	48.4853
	7 × 7	17.3403	26.8960	42.5587	44.1002	51.6061	84.6406	145.0541
	9 × 9	17.3403	26.8874	42.4727	44.1002	51.6061	66.8809	84.6406
2/3	11 × 11	17.3400	26.7827	42.4727	44.1002	51.6061	66.8809	70.2020
	13 × 13	17.3371	26.7827	42.4727	44.0330	51.2546	66.8809	70.2020
	14 × 14	17.3371	26.7827	42.4727	44.0330	51.2543	66.8809	70.2020
	7 × 7	23.1091	47.2134	47.6925	69.7066	88.1315	89.0025	148.6571
	9 × 9	23.1091	47.1380	47.6203	69.7066	88.1315	89.0025	108.3649
1.0	11 × 11	23.1088	47.1380	47.1380	69.7066	84.9829	88.6099	108.3649
	13 × 13	23.1023	47.1380	47.1380	69.6014	84.8250	88.3922	108.3649
	14 × 14	23.1023	47.1380	47.1380	69.4976	84.8250	88.3922	108.3649
	7 × 7	39.0157	60.2816	97.1983	99.2255	116.1137	157.9645	190.4414
	9 × 9	39.0157	60.2610	97.0106	99.2255	116.1137	150.4820	157.9546
1.5	11 × 11	39.0147	60.2610	95.5637	96.3840	116.1137	150.4820	157.9546
	13 × 13	39.0080	60.2610	95.5637	96.2255	116.1133	150.4820	157.9546
	14 × 14	39.0080	60.2610	95.5637	96.2255	115.3222	150.4820	157.9546

Source: Reproduced from Chakraverty, S., Pradhan, K.K., Free vibration of exponential functionally graded rectangular plates in thermal environment with general boundary conditions. Aerospace Science and Technology 2014;36:132–156.

Table 5.6 Comparison of the first seven non-dimensional frequencies for isotropic square plates

BCs	Sources	λ_1	λ_2	λ_3	λ_4	λ_5	λ_6	λ_7
CCCC	Present	35.9888	73.3989	73.3989	108.2653	131.8982	132.4236	165.2066
	Leissa (1973)	35.990	73.390	73.390	108.220	131.580	132.200	165.000
	Singh and Chakraverty (1994a)	35.988	73.398	73.398	108.26	131.89	-	-
	Liu and Liew (1999)	35.9375	73.2324	73.2324	107.8890	131.1188	131.7522	-
	Yang and Shen (2001)	35.9879	73.4053	73.4172	108.2728	131.1693	132.0165	-
	Bhat (1985)	35.9855	73.395	73.395	108.22	131.78	132.41	-
	Liew et al. (1993)	35.9875	73.3943	73.3943	108.2172	131.5766	132.2043	-
	Geannakakes (1995)	35.9852	73.3939	73.3939	108.2166	131.5823	132.2070	165.0019
	Cheung and Zhou (1999)	36.004	73.432	73.432	108.27	131.83	-	-
CCCS	Present	31.8278	63.3399	71.0835	100.8305	116.5434	130.6136	153.8345
	Leissa (1973)	31.829	63.347	71.084	100.830	116.400	130.370	-
	Singh and Chakraverty (1994a)	31.827	63.339	71.098	100.83	116.54	-	-
	Yang and Shen (2001)	31.8237	63.3391	71.0829	100.8310	116.2164	130.3377	-
	Liew et al. (1993)	31.8285	63.3322	71.0760	100.7914	116.3557	130.3479	-
CSCS	Present	28.9509	54.7439	69.3271	94.7034	103.7178	129.3120	141.3810
	Leissa (1973)	28.9505	54.7431	69.3270	94.5853	102.2162	129.0955	-
	Singh and Chakraverty (1994a)	28.950	54.873	69.327	94.703	103.71	-	-
	Liu and Liew (1999)	28.9216	54.6658	69.1927	94.3594	101.9944	128.6742	-
	Yang and Shen (2001)	28.9460	54.7410	69.3303	94.6120	102.1651	129.0791	-
	Liew et al. (1993)	28.9515	54.7418	69.3261	94.5834	102.2136	129.0915	-
SSSS	Present	19.7392	49.3490	49.3490	79.4007	100.1729	100.1868	130.3895
	Singh and Chakraverty (1994a)	19.739	49.348	49.348	79.400	100.17	-	-
	Bhat (1985)	19.739	49.348	49.348	78.957	99.304	99.304	-
	Cheung and Zhou (1999)	19.743	49.354	49.354	78.971	98.733	-	-

Source: First published in Chakraverty, S., Pradhan, K.K., 2014. Free vibration of functionally graded thin rectangular plates resting on Winkler elastic foundation with general boundary conditions using Rayleigh-Ritz method. Int. J. Appl. Mech. 06, 1450043.

first seven non-dimensional frequencies of FG square Al/Al_2O_3 plate with the available literature is performed. In a similar fashion, FG rectangular $SUS304/Si_3N_4$ plates with different edge ratios are considered in Table 5.7. It is evident from Tables 5.6 and 5.7 that present results for non-dimensional frequencies are in excellent agreement with open literature.

5.3 RESULTS AND DISCUSSIONS

In this domain, free vibration frequencies of power-law variation based FG plate have been evaluated first and subsequently the exponential gradation is assumed for such computation.

5.3.1 Power-Law Gradation

First of all, let us evaluate the natural frequencies of FG rectangular plates with power-law gradation of material properties. Looking into the convergence and comparison of non-dimensional frequencies, the results for free vibration of FG rectangular plates for different combinations of BCs with different k and a/b are reported. Consequently, natural frequencies of clamped FG rectangular plates are reported in Table 5.8. In a similar manner, non-dimensional frequencies of FG plates subject to different sets of BCs are evaluated in Tables 5.9–5.14. It may be noticed that non-dimensional frequencies are increasing with an increase in aspect ratios for a fixed power-law index irrespective of edge support and geometric configuration considered, whereas one may observe descending pattern of present results with an increase in power-law indices for a fixed aspect ratio. The non-dimensional frequencies for other combinations of edge supports can also be found after analyzing these results.

After observing Tables 5.8–5.14, one may plot the effect of aspect ratios (a/b) and power-law indices (k) on non-dimensional frequencies of FG rectangular plates under any combination of BCs. Figure 5.1 depicts the variation of non-dimensional frequencies for clamped rectangular plates with respect to these parameters. It is evident from Fig. 5.1 that non-dimensional frequencies go on increasing with an increase in aspect ratios for a fixed power-law index. It is also interesting to note that frequencies are decreasing with an increase in power-law indices for a fixed aspect ratio. Similar behavior may also be observed for other BCs.

The first six mode shapes of FG square plates subject to various BCs, viz. CCCC, CCCS, CCCF, and SSSS, are given in Figs. 5.2–5.5 with $k = 0$. With similar BCs, Figs. 5.6–5.9 depict the first six mode shapes with

Table 5.7 Comparison of the first seven non-dimensional frequencies for CCCC isotropic rectangular plates

a/b	Sources	λ_1	λ_2	λ_3	λ_4	λ_5	λ_6	λ_7
2/5	Present	23.6442	27.8095	35.4201	46.8183	62.0985	63.0861	67.3955
	Leissa (1973)	23.648	27.817	35.446	46.702	61.554	63.100	-
	Yang and Shen (2001)	23.6530	27.8243	35.4280	46.7023	61.5044	63.1151	67.4557
	Liew et al. (1993)	23.6428	27.8056	35.4158	46.6687	61.5091	63.0758	67.3837
	Geannakakes (1995)	23.6438	27.8070	35.4179	46.6762	62.0651	63.0829	67.3906
2/3	Present	27.0075	41.7073	66.1280	66.6235	79.8454	101.3552	103.2086
	Leissa (1973)	27.010	41.716	66.143	66.552	79.850	100.850	-
	Yang and Shen (2001)	27.0114	41.7184	66.1516	66.5187	79.8664	100.8301	103.1675
	Geannakakes (1995)	27.0050	41.7038	66.1245	66.5225	79.8051	100.8273	103.1261
1.5	Present	60.7626	93.8415	148.7881	149.6842	179.5755	228.0491	232.2193
	Leissa (1973)	60.772	93.860	148.820	149.740	179.660	226.92	-
	Yang and Shen (2001)	60.7527	93.8398	148.7879	149.7063	179.6004	226.9138	232.1345
	Bhat (1985)	60.962	93.835	148.78	149.85	179.57	227.90	-
	Liew et al. (1993)	60.7662	93.8390	148.7798	149.6770	179.5649	226.8228	232.0324

Source: First published in Chakraverty, S., Pradhan, K.K., 2014. Free vibration of functionally graded thin rectangular plates resting on Winkler elastic foundation with general boundary conditions using Rayleigh-Ritz method. Int. J. Appl. Mech. 06, 1450043.

Table 5.8 Effect of aspect ratios on non-dimensional frequencies of CCCC FG rectangular plates with different *k*

k	*a/b*	λ_1	λ_2	λ_3	λ_4	λ_5	λ_6
0.0	0.2	22.633	23.440	24.877	27.039	30.818	36.279
	0.5	24.579	31.829	44.819	63.598	63.986	71.119
	1.0	35.989	73.399	73.399	108.27	131.89	132.42
	2.0	98.318	127.32	179.28	254.39	255.95	284.48
0.2	0.2	21.176	21.931	23.275	25.298	28.834	33.943
	0.5	22.997	29.779	41.933	59.503	59.866	66.540
	1.0	33.671	68.673	68.673	101.29	123.41	123.89
	2.0	91.987	119.12	167.73	238.01	239.47	266.16
0.5	0.2	19.879	20.588	21.849	23.748	27.068	31.864
	0.5	21.588	27.956	39.365	55.858	56.199	62.465
	1.0	31.609	64.467	64.467	95.091	115.85	116.31
	2.0	86.354	111.82	157.46	223.43	224.80	249.86
1.0	0.2	18.832	19.503	20.699	22.498	25.642	30.186
	0.5	20.451	26.484	37.292	52.916	53.239	59.175
	1.0	29.945	61.072	61.072	90.082	109.75	110.18
	2.0	81.805	105.93	149.17	211.67	212.96	236.7
2.0	0.2	18.002	18.643	19.786	21.506	24.512	28.855
	0.5	19.549	25.316	35.648	50.583	50.893	56.566
	1.0	28.624	58.379	58.379	86.111	104.91	105.33
	2.0	78.199	101.26	142.59	202.33	203.57	226.26

Source: First published in Chakraverty, S., Pradhan, K.K., 2014. Free vibration of functionally graded thin rectangular plates resting on Winkler elastic foundation with general boundary conditions using Rayleigh-Ritz method. Int. J. Appl. Mech. 06, 1450043.

Table 5.9 Effect of aspect ratios on non-dimensional frequencies of CCCS FG rectangular plates with different *k*

k	*a/b*	λ_1	λ_2	λ_3	λ_4	λ_5	λ_6
0.0	0.2	15.771	16.849	18.704	21.458	25.894	35.914
	0.5	18.349	27.056	41.300	52.643	60.693	61.341
	1.0	31.828	63.339	71.084	100.83	116.54	130.61
	2.0	96.574	121.01	167.13	244.24	254.99	280.71
0.2	0.2	14.756	15.765	17.499	20.077	24.227	33.601
	0.5	17.168	25.314	38.640	49.253	56.785	57.392
	1.0	29.778	59.262	66.506	94.338	109.04	122.20
	2.0	90.356	113.22	156.37	228.51	238.58	262.64
0.5	0.2	13.852	14.799	16.428	18.847	22.743	31.543
	0.5	16.116	23.764	36.274	46.236	53.307	53.877
	1.0	27.955	55.632	62.433	88.560	102.36	114.72
	2.0	84.822	106.28	146.79	214.51	223.97	246.55

Continued

Table 5.9 Effect of aspect ratios on non-dimensional frequencies of CCCS FG rectangular plates with different *k*—cont'd

k	a/b	λ_1	λ_2	λ_3	λ_4	λ_5	λ_6
1.0	0.2	13.122	14.019	15.563	17.854	21.545	29.882
	0.5	15.268	22.512	34.363	43.801	50.499	51.039
	1.0	26.482	52.702	59.145	83.896	96.969	108.68
	2.0	80.355	100.69	139.06	203.22	212.17	233.57
2.0	0.2	12.544	13.402	14.877	17.067	20.595	28.565
	0.5	14.595	21.520	32.848	41.870	48.273	48.789
	1.0	25.315	50.379	56.538	80.197	92.695	103.89
	2.0	76.812	96.247	132.93	194.26	202.82	223.27

Source: First published in Chakraverty, S., Pradhan, K.K., 2014. Free vibration of functionally graded thin rectangular plates resting on Winkler elastic foundation with general boundary conditions using Rayleigh-Ritz method. Int. J. Appl. Mech. 06, 1450043.

Table 5.10 Effect of aspect ratios on non-dimensional frequencies of CFCF FG rectangular plates with different *k*

k	a/b	λ_1	λ_2	λ_3	λ_4	λ_5	λ_6
0.0	0.2	0.8771	2.4095	2.8869	4.7523	5.9772	7.9153
	0.5	5.5287	9.0046	15.224	20.709	27.663	30.068
	1.0	22.224	26.478	44.073	61.265	67.649	82.254
	2.0	89.179	93.709	111.01	145.61	245.99	252.95
0.2	0.2	0.8206	2.2543	2.7011	4.4463	5.5924	7.4057
	0.5	5.1727	8.4248	14.244	19.376	25.881	28.132
	1.0	20.793	24.773	41.235	57.321	63.294	76.958
	2.0	83.437	87.676	103.86	136.24	230.15	236.67
0.5	0.2	0.7703	2.1163	2.5356	4.1740	5.2499	6.9521
	0.5	4.8559	7.9088	13.372	18.189	24.296	26.409
	1.0	19.519	23.256	38.709	53.810	59.417	72.244
	2.0	78.327	82.306	97.503	127.89	216.06	222.17
1.0	0.2	0.7298	2.0048	2.4021	3.9542	4.9734	6.5859
	0.5	4.6002	7.4923	12.667	17.231	23.017	25.018
	1.0	18.491	22.031	36.671	50.976	56.288	68.439
	2.0	74.202	77.971	92.367	121.16	204.68	210.47
2.0	0.2	0.6976	1.9164	2.2962	3.7798	4.7541	6.2956
	0.5	4.3973	7.1619	12.109	16.471	22.002	23.915
	1.0	17.676	21.059	35.054	48.729	53.806	65.422
	2.0	70.930	74.533	88.295	115.81	195.66	201.19

Source: First published in Chakraverty, S., Pradhan, K.K., 2014. Free vibration of functionally graded thin rectangular plates resting on Winkler elastic foundation with general boundary conditions using Rayleigh-Ritz method. Int. J. Appl. Mech. 06, 1450043.

Table 5.11 Effect of aspect ratios on non-dimensional frequencies of SCSC FG rectangular plates with different *k*

k	*a/b*	λ_1	λ_2	λ_3	λ_4	λ_5	λ_6
0.0	0.2	22.593	23.277	24.521	26.479	35.417	47.242
	0.5	23.816	28.951	39.351	56.145	63.535	69.373
	1.0	28.951	54.744	69.327	94.703	103.72	129.31
	2.0	54.743	94.585	154.97	170.36	207.34	235.68
0.2	0.2	21.138	21.779	22.942	24.775	33.136	44.199
	0.5	22.282	27.087	36.817	52.529	59.444	64.906
	1.0	27.087	51.219	64.863	88.605	97.039	120.99
	2.0	51.218	88.495	144.99	159.39	193.99	220.51
0.5	0.2	19.843	20.445	21.537	23.258	31.107	41.493
	0.5	20.918	25.428	34.562	49.312	55.803	60.931
	1.0	25.428	48.082	60.891	83.179	91.096	113.58
	2.0	48.081	83.075	136.11	149.63	182.11	207.00
1.0	0.2	18.798	19.368	20.403	22.033	29.469	39.308
	0.5	19.816	24.089	32.742	46.715	52.864	57.722
	1.0	24.089	45.549	57.684	78.798	86.298	107.59
	2.0	45.549	78.699	128.94	141.75	172.51	196.10
2.0	0.2	17.969	18.514	19.503	21.061	28.169	37.575
	0.5	18.942	23.027	31.298	44.656	50.533	55.177
	1.0	23.027	43.542	55.140	75.324	82.494	102.85
	2.0	43.541	75.230	123.26	135.50	164.91	187.45

Source: First published in Chakraverty, S., Pradhan, K.K., 2014. Free vibration of functionally graded thin rectangular plates resting on Winkler elastic foundation with general boundary conditions using Rayleigh-Ritz method. Int. J. Appl. Mech. 06, 1450043.

Table 5.12 Effect of aspect ratios on non-dimensional frequencies of SSSS FG rectangular plates with different *k*

k	*a/b*	λ_1	λ_2	λ_3	λ_4	λ_5	λ_6
0.0	0.2	10.264	11.449	13.495	16.433	28.514	39.874
	0.5	12.337	19.739	32.421	41.947	49.659	50.916
	1.0	19.739	49.349	49.349	79.401	100.17	100.19
	2.0	49.348	78.958	129.68	167.79	198.63	203.66
0.2	0.2	9.6035	10.712	12.626	15.374	26.678	37.307
	0.5	11.543	18.468	30.333	39.246	46.461	47.637
	1.0	18.468	46.171	46.171	74.288	93.723	93.736
	2.0	46.171	73.874	121.33	156.98	185.84	190.55
0.5	0.2	9.0153	10.056	11.853	14.433	25.044	35.022
	0.5	10.836	17.337	28.476	36.842	43.616	44.719
	1.0	17.337	43.344	43.344	69.738	87.983	87.995
	2.0	43.343	69.349	113.90	147.37	174.46	178.88

Continued

Table 5.12 Effect of aspect ratios on non-dimensional frequencies of SSSS FG rectangular plates with different *k*—cont'd

k	a/b	λ_1	λ_2	λ_3	λ_4	λ_5	λ_6
1.0	0.2	8.5405	9.5260	11.228	13.673	23.725	33.177
	0.5	10.265	16.424	26.976	34.902	41.319	42.364
	1.0	16.424	41.061	41.061	66.065	83.349	83.360
	2.0	41.060	65.697	107.90	139.61	165.27	169.46
2.0	0.2	8.1639	9.1060	10.733	13.069	22.679	31.715
	0.5	9.8125	15.700	25.787	33.363	39.497	40.497
	1.0	15.699	39.251	39.251	63.153	79.674	79.685
	2.0	39.249	62.800	103.15	133.45	157.99	161.99

Source: First published in Chakraverty, S., Pradhan, K.K., 2014. Free vibration of functionally graded thin rectangular plates resting on Winkler elastic foundation with general boundary conditions using Rayleigh-Ritz method. Int. J. Appl. Mech. 06, 1450043.

Table 5.13 Effect of aspect ratios on non-dimensional frequencies of CCCF FG rectangular plates with different *k*

k	a/b	λ_1	λ_2	λ_3	λ_4	λ_5	λ_6
0.0	0.2	3.9902	5.3684	7.6058	10.729	16.579	22.576
	0.5	7.7876	17.559	25.856	32.359	36.053	51.535
	1.0	23.961	40.019	63.291	76.759	80.689	117.86
	2.0	90.659	104.17	135.11	192.94	247.72	264.23
0.2	0.2	3.7332	5.0227	7.1161	10.039	15.511	21.123
	0.5	7.2861	16.428	24.191	30.275	33.732	48.216
	1.0	22.418	37.443	59.216	71.816	75.494	110.27
	2.0	84.821	97.465	126.41	180.52	231.77	247.22
0.5	0.2	3.5046	4.7151	6.6802	9.4242	14.561	19.829
	0.5	6.8399	15.422	22.709	28.421	31.666	45.263
	1.0	21.045	35.149	55.589	67.418	70.870	103.52
	2.0	79.627	91.496	118.67	169.46	217.57	232.08
1.0	0.2	3.3200	4.4668	6.3284	8.9278	13.794	18.785
	0.5	6.4796	14.609	21.513	26.924	29.998	42.879
	1.0	19.936	33.298	52.662	63.867	67.138	98.063
	2.0	75.433	86.677	112.42	160.54	206.11	219.85
2.0	0.2	3.1737	4.2699	6.0494	8.5342	13.186	17.957
	0.5	6.1939	13.966	20.565	25.737	28.676	40.989
	1.0	19.057	31.830	50.339	61.051	64.178	93.740
	2.0	72.107	82.856	107.46	153.46	197.03	210.16

Source: First published in Chakraverty, S., Pradhan, K.K., 2014. Free vibration of functionally graded thin rectangular plates resting on Winkler elastic foundation with general boundary conditions using Rayleigh-Ritz method. Int. J. Appl. Mech. 06, 1450043.

Table 5.14 Effect of aspect ratios on non-dimensional frequencies of CSCF FG rectangular plates with different k

k	a/b	λ_1	λ_2	λ_3	λ_4	λ_5	λ_6
0.0	0.2	1.6647	3.6854	6.2933	9.6409	15.569	16.067
	0.5	6.6129	16.819	19.958	31.184	31.795	47.634
	1.0	23.412	35.595	62.961	67.356	77.540	111.42
	2.0	90.418	101.66	128.85	182.05	247.69	262.47
0.2	0.2	1.5575	3.4481	5.8880	9.0202	14.567	15.033
	0.5	6.1871	15.736	18.673	29.176	29.748	44.567
	1.0	21.904	33.303	58.907	63.019	72.548	104.24
	2.0	84.596	95.113	120.55	170.33	23.174	245.57
0.5	0.2	14.621	3.2369	5.5274	8.4678	13.675	14.112
	0.5	5.8082	14.772	17.529	27.389	27.926	41.837
	1.0	20.563	31.264	55.299	59.159	68.104	97.857
	2.0	79.415	89.288	113.17	159.89	217.55	230.53
1.0	0.2	1.3851	3.0665	5.2363	8.0218	12.955	13.369
	0.5	5.5023	13.994	16.606	25.947	26.455	39.634
	1.0	19.479	29.617	52.386	56.043	64.517	92.703
	2.0	75.232	84.585	107.21	151.47	206.09	218.39
2.0	0.2	1.3240	2.9313	5.0055	7.6681	12.383	12.779
	0.5	5.2598	13.377	15.874	24.803	25.289	37.886
	1.0	18.621	28.311	50.077	53.572	61.673	88.616
	2.0	71.915	80.856	102.48	144.79	197.00	208.76

Source: First published in Chakraverty, S., Pradhan, K.K., 2014. Free vibration of functionally graded thin rectangular plates resting on Winkler elastic foundation with general boundary conditions using Rayleigh-Ritz method. Int. J. Appl. Mech. 06, 1450043.

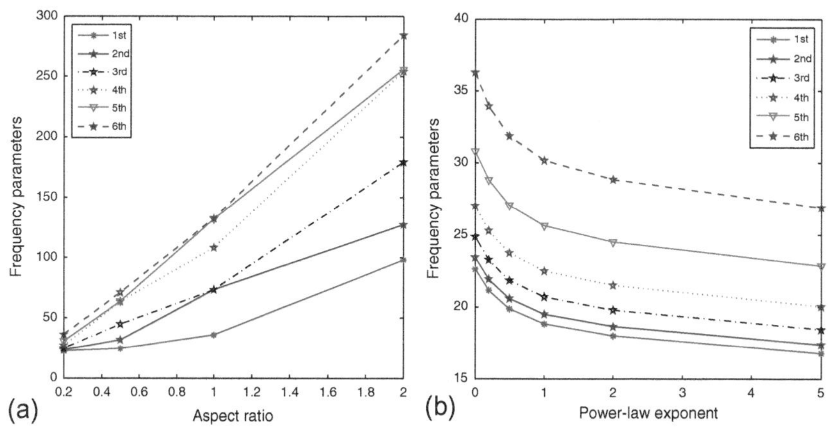

Figure 5.1 Effect of (a) aspect ratios and (b) power-law indices on non-dimensional frequencies of CCCC FG rectangular plate. (*Source: First published in Chakraverty, S., Pradhan, K.K., 2014. Free vibration of functionally graded thin rectangular plates resting on Winkler elastic foundation with general boundary conditions using Rayleigh-Ritz method. Int. J. Appl. Mech. 06, 1450043*).

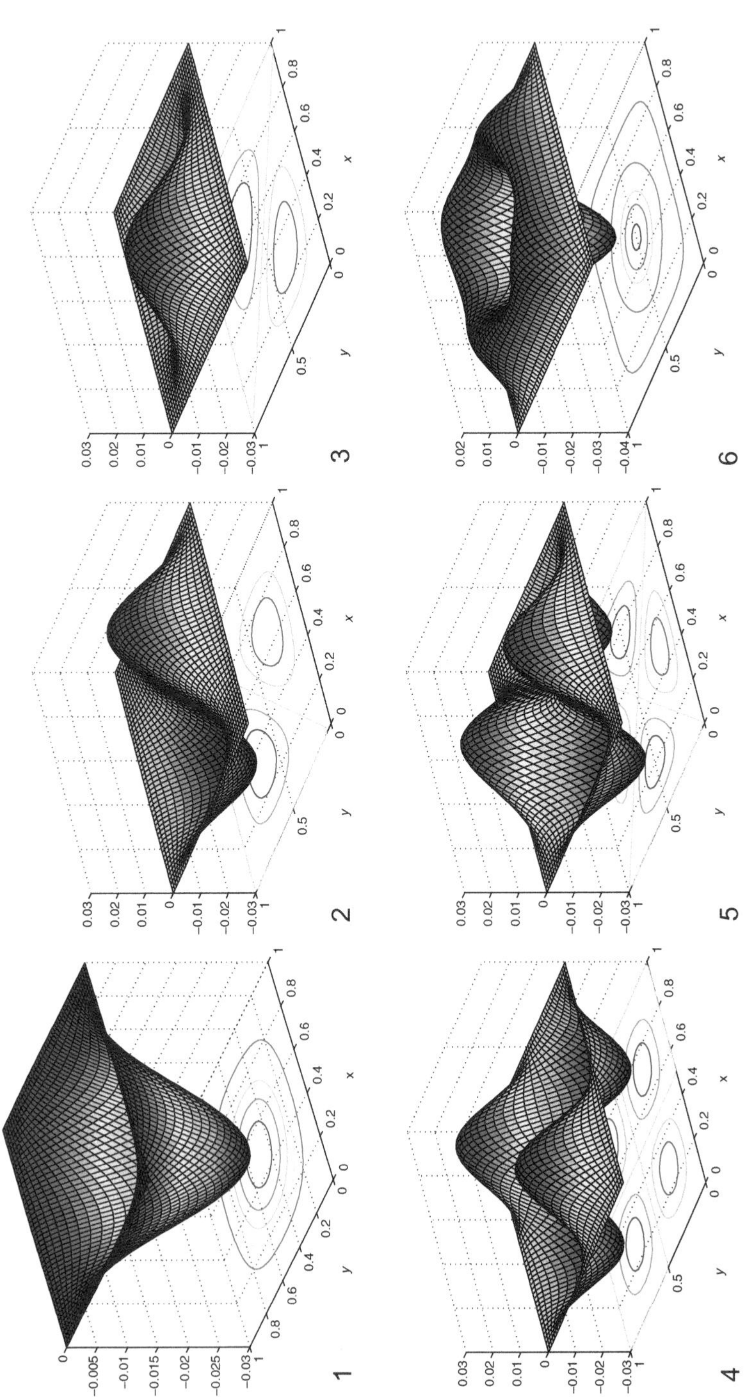

Figure 5.2 The first six mode shapes of CCCC FG square plate with $k = 0$. (*Source: First published in Chakraverty, S., Pradhan, K.K., 2014. Free vibration of functionally graded thin rectangular plates resting on Winkler elastic foundation with general boundary conditions using Rayleigh-Ritz method. Int. J. Appl. Mech. 06, 1450043*).

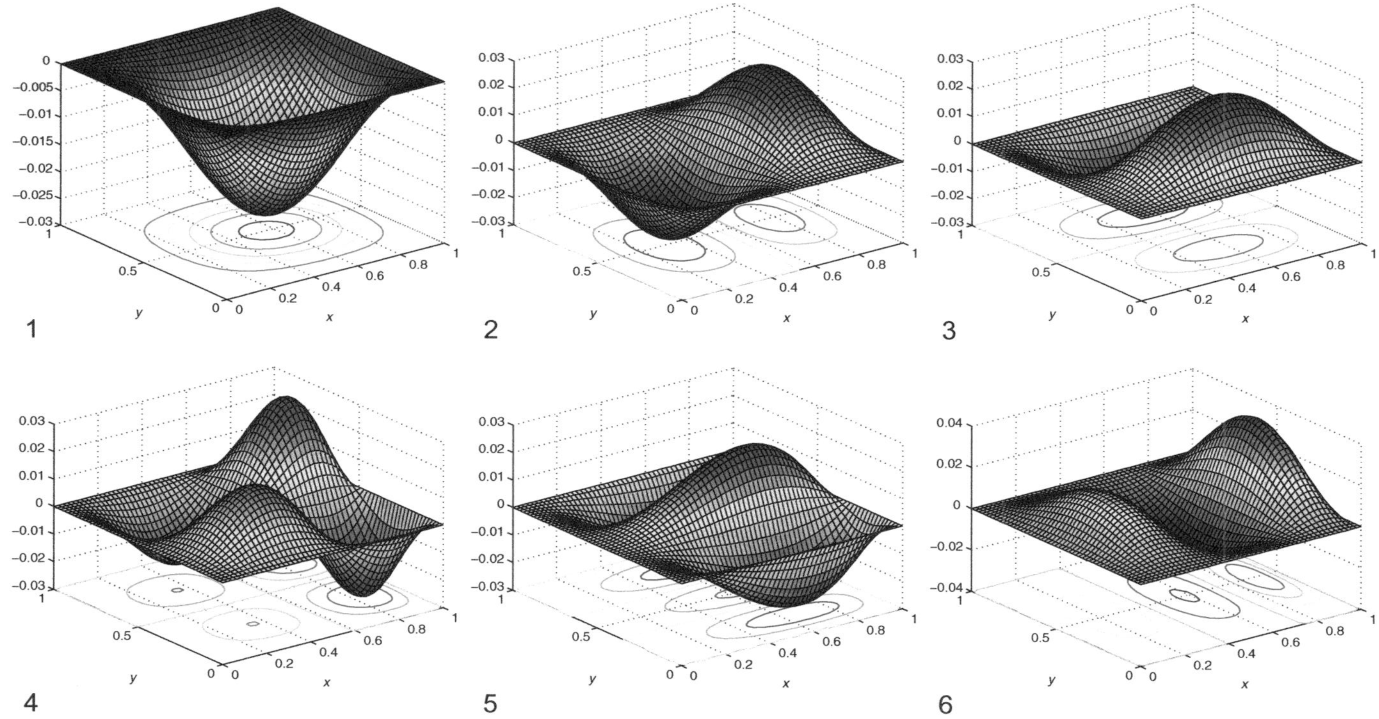

Figure 5.3 The first six mode shapes of CCCS FG square plate with $k = 0$. (*Source: First published in Chakraverty, S., Pradhan, K.K., 2014. Free vibration of functionally graded thin rectangular plates resting on Winkler elastic foundation with general boundary conditions using Rayleigh-Ritz method. Int. J. Appl. Mech. 06, 1450043*).

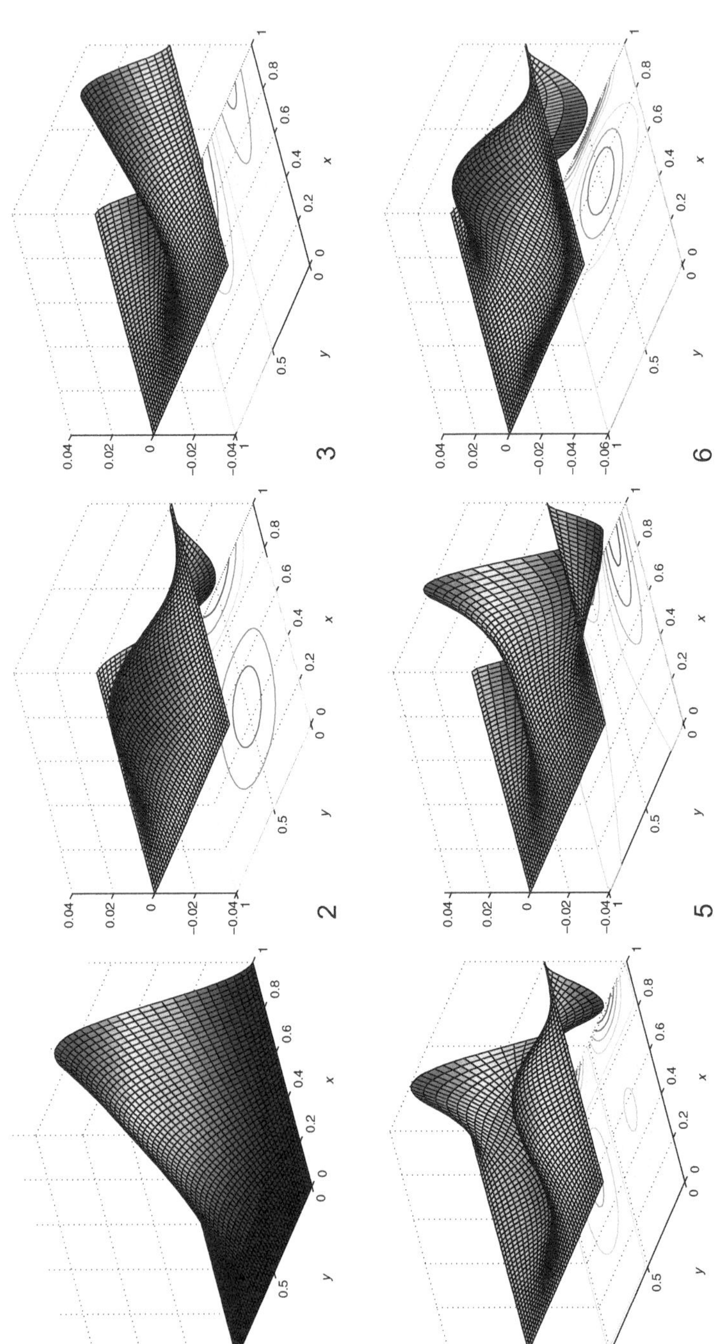

Figure 5.4 The first six mode shapes of CCCF FG square plate with $k = 0$. (*Source: First published in Chakraverty, S., Pradhan, K.K., 2014. Free vibration of functionally graded thin rectangular plates resting on Winkler elastic foundation with general boundary conditions using Rayleigh-Ritz method. Int. J. Appl. Mech. 06, 1450043*).

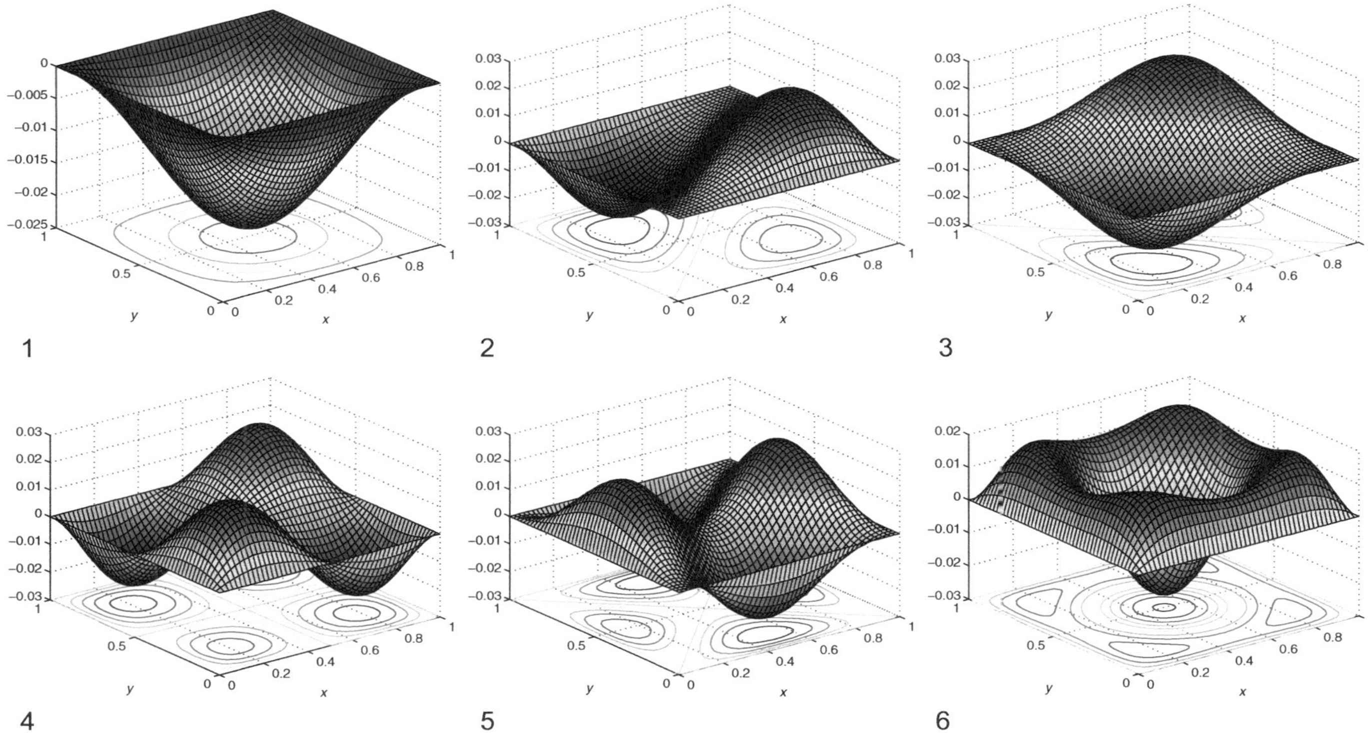

Figure 5.5 The first six mode shapes of SSSS FG square plate with $k = 0$. (*Source: First published in Chakraverty, S., Pradhan, K.K., 2014. Free vibration of functionally graded thin rectangular plates resting on Winkler elastic foundation with general boundary conditions using Rayleigh-Ritz method. Int. J. Appl. Mech. 06, 1450043*).

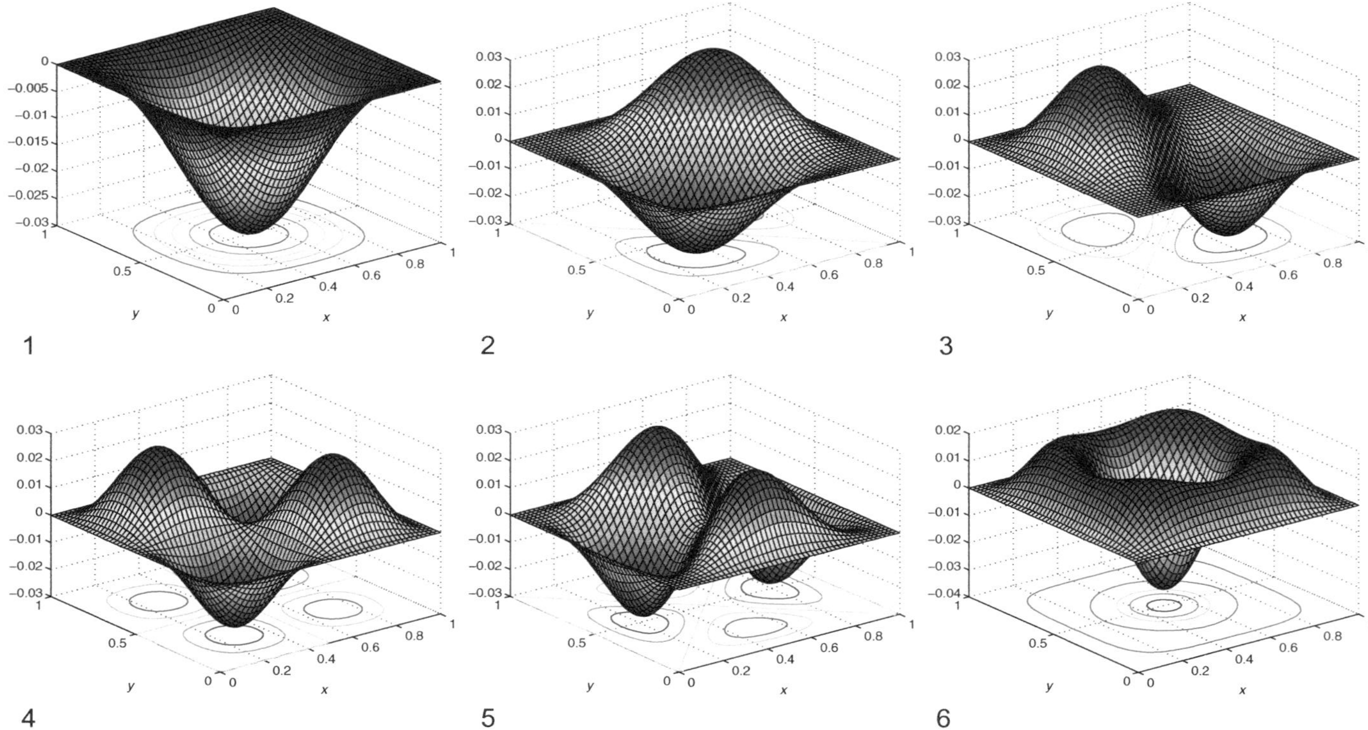

Figure 5.6 The first six mode shapes of CCCC FG square plate with $k = 0.5$. (*Source: First published in Chakraverty, S., Pradhan, K.K., 2014. Free vibration of functionally graded thin rectangular plates resting on Winkler elastic foundation with general boundary conditions using Rayleigh-Ritz method. Int. J. Appl. Mech. 06, 1450043*).

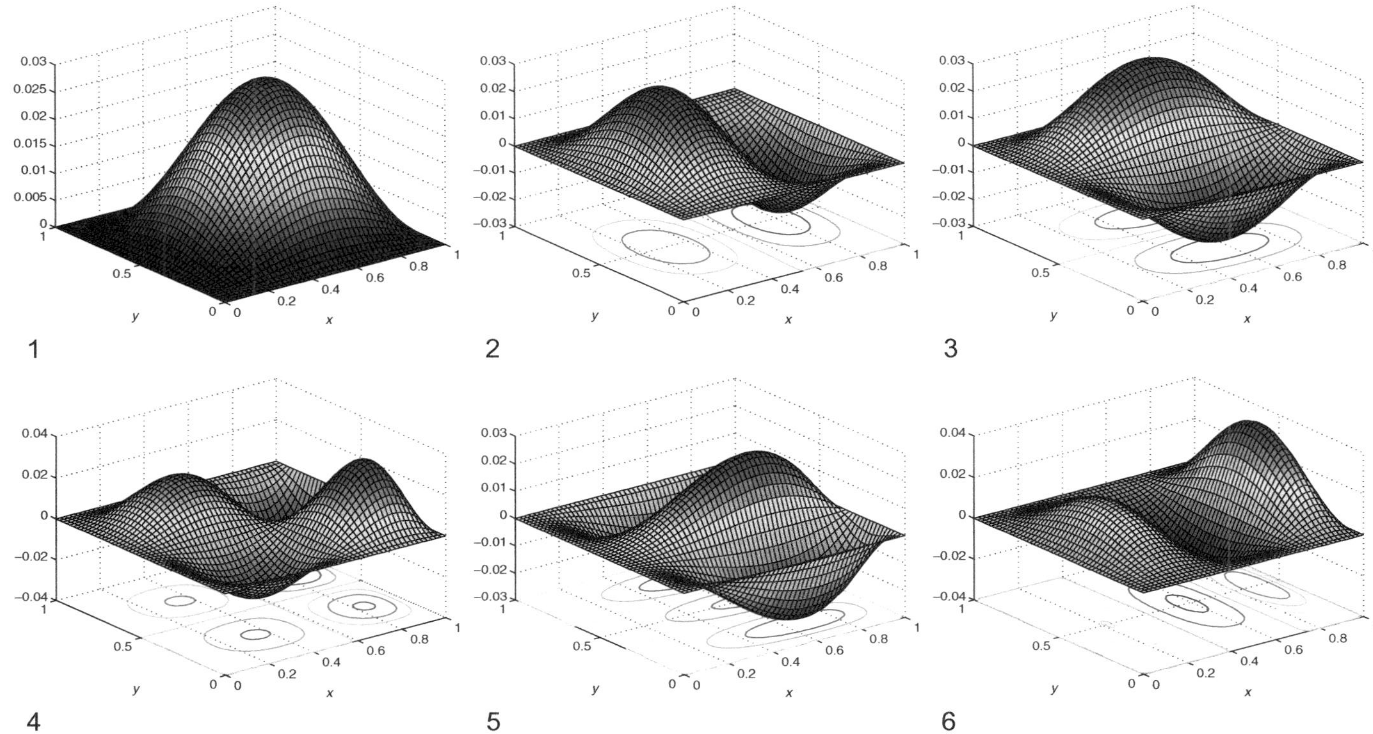

Figure 5.7 The first six mode shapes of CCCS FG square plate with $k = 0.5$. (*Source: First published in Chakraverty, S., Pradhan, K.K., 2014. Free vibration of functionally graded thin rectangular plates resting on Winkler elastic foundation with general boundary conditions using Rayleigh-Ritz method. Int. J. Appl. Mech. 06, 1450043*).

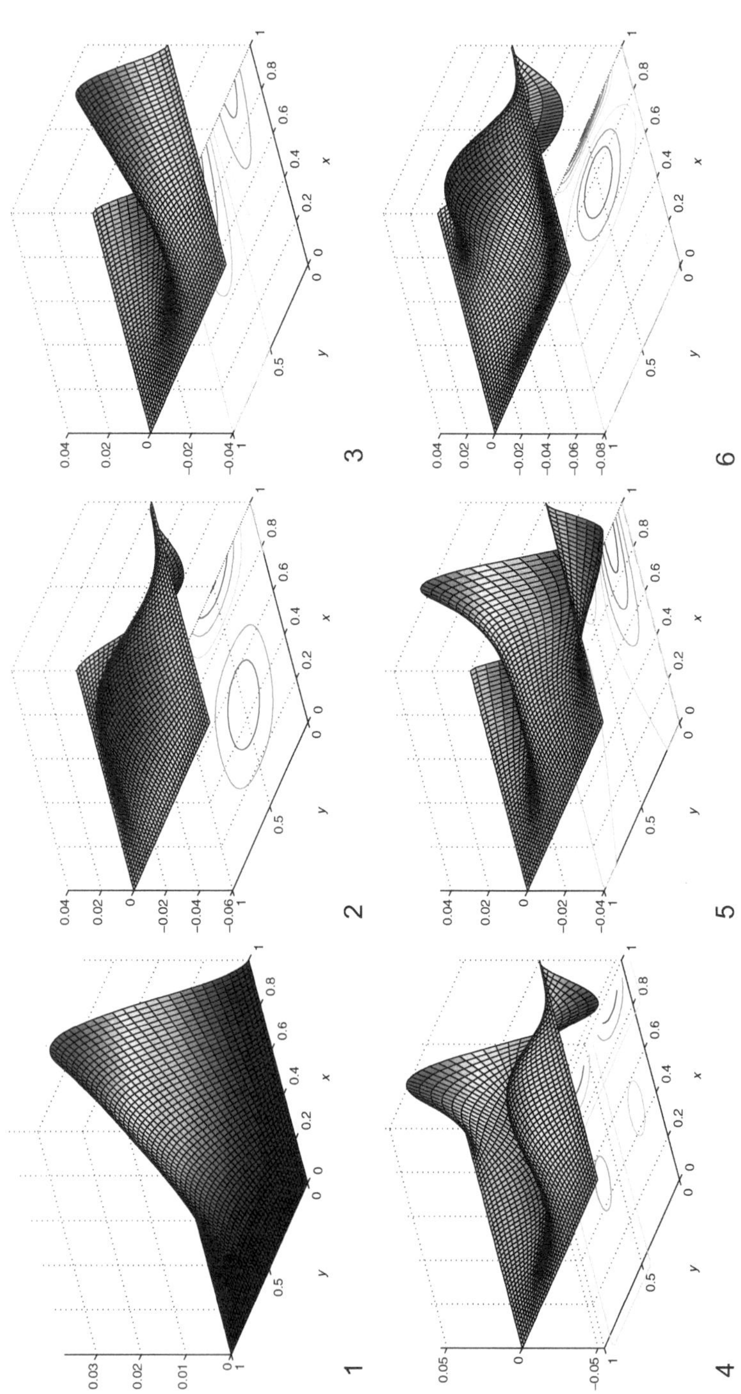

Figure 5.8 The first six mode shapes of CCCF FG square plate with $k = 0.5$. (*Source: First published in Chakraverty, S., Pradhan, K.K., 2014. Free vibration of functionally graded thin rectangular plates resting on Winkler elastic foundation with general boundary conditions using Rayleigh-Ritz method. Int. J. Appl. Mech. 06, 1450043*).

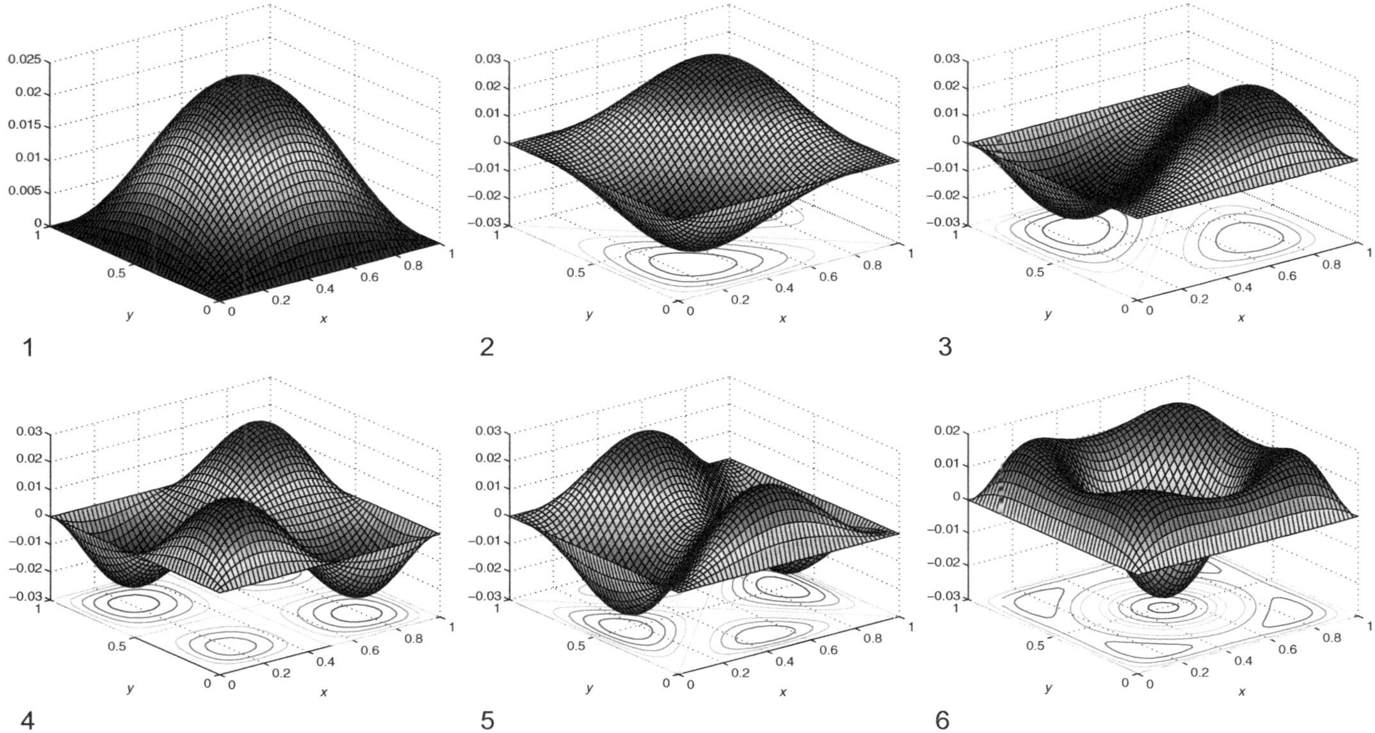

Figure 5.9 The first six mode shapes of SSSS FG square plate with $k = 0.5$. (*Source: First published in Chakraverty, S., Pradhan, K.K., 2014. Free vibration of functionally graded thin rectangular plates resting on Winkler elastic foundation with general boundary conditions using Rayleigh-Ritz method. Int. J. Appl. Mech. 06, 1450043*).

$k = 0.5$. In a similar way, one may also see the deflected shapes for other sets of BCs.

5.3.2 Exponential Gradation

The natural frequencies of exponential FG plate with 24 possible combinations of boundary conditions have been computed in Table 5.15 with different edge (aspect) ratios, viz. $a/b = 0.2, 0.5, 1.0, 2.0, 2.5$. When analyzing results obtained for different edge supports, it may be observed that the results are increasing with an increase in aspect ratios. This fact may be true because the FG rectangular plate is getting gradually flexible with an increase in aspect ratios.

The first six mode shapes of clamped FG rectangular plate are given in Figs. 5.10 and 5.11 with different aspect ratios, viz. $a/b = 0.2, 0.5, 1.0$, and 2.0. Deflections of 3D mode shapes are depicted in different directions with the variation of a/b. Similarly, 3D mode shapes for other boundary supports can also be presented.

5.3.2.1 *Effect of* E_{rat} $(= E_c/E_m)$ *With Constant* ρ_{rat} $(= \rho_c/\rho_m)$

In Tables 5.16–5.21, the effects of E_{rat} on frequencies of exponential FG rectangular plates subject to various edge supports are tabulated with constant $\rho_{rat} = 1$. In Table 5.16, frequencies of FG plates with the variation of E_{rat}, viz. $E_{rat} = 0.25, 0.5, 1.0$, and 2.0, are evaluated with a fixed aspect ratio. In a similar fashion, the effect of E_{rat} on the frequencies are also computed in Tables 5.17–5.21 with different BCs, viz. CCCS, CCSS, CCCF, SSSS, and FFFF. It is worth mentioning that natural frequencies are decreasing with an increase in E_{rat} irrespective of aspect ratio and BC considered. It can be found in the expression of natural frequency that it is inversely proportional to the Young's modulus of the ceramic constituent of FG rectangular plates and hence this behavior with an increase in E_{rat} may in fact be true.

Accordingly, the first six mode shapes of CCCC FG square plates ($a/b = 1$) with different E_{rat}, viz. $E_{rat} = 0.25, 0.5, 1.0$, and 2.0, are demonstrated in Figs. 5.12 and 5.13. It is interesting to note here that the respective 3D mode shapes are quite similar, but deflections occur in different directions for higher modes.

5.3.2.2 *Effect of* ρ_{rat} *With Constant* E_{rat}

In Tables 5.22–5.27, the effect of ρ_{rat} on frequencies of exponential FG rectangular plates satisfying various BCs are estimated with constant

Table 5.15 Effect of aspect ratios on non-dimensional frequencies of exponential FG plate with different combinations of boundary conditions

BCs	a/b	Ceramic constituent of Al/Al_2O_3 ($\delta = 0$)						Exponential Al/Al_2O_3 FG plate					
		λ_1	λ_2	λ_3	λ_4	λ_5	λ_6	λ_1	λ_2	λ_3	λ_4	λ_5	λ_6
CCCC	0.2	17.824	18.459	19.591	21.293	24.269	28.570	15.024	15.559	16.514	17.949	20.458	24.083
	0.5	19.357	25.066	35.296	50.084	50.390	56.008	16.316	21.129	29.752	42.217	42.475	47.210
	1.0	28.342	57.803	57.803	85.261	103.87	104.29	23.890	48.724	48.724	71.868	87.556	87.905
	2.0	77.427	100.26	141.18	200.34	201.56	224.03	65.265	84.515	119.01	168.87	169.90	188.84
	2.5	116.38	136.88	174.46	230.44	310.51	331.93	98.099	115.38	147.06	194.24	261.74	279.79
CCCS	0.2	12.420	13.269	14.729	16.899	20.392	28.283	10.469	11.185	12.416	14.244	17.189	23.840
	0.5	14.450	21.307	32.524	41.457	47.796	48.307	12.181	17.961	27.416	34.945	40.289	40.719
	1.0	25.065	49.881	55.979	79.405	91.779	102.86	21.128	42.046	47.187	66.933	77.364	86.704
	2.0	76.054	95.297	131.62	192.34	200.81	221.07	64.108	80.328	110.94	162.13	169.27	186.34
	2.5	115.37	132.98	166.38	223.12	309.95	310.47	97.247	112.09	140.25	188.07	261.27	261.71
CCSS	0.2	12.397	13.178	14.533	16.626	21.072	39.639	10.449	11.108	12.250	14.015	17.762	33.413
	0.5	13.994	19.846	29.969	41.232	46.022	46.987	11.795	16.729	25.262	34.756	38.793	39.607
	1.0	21.306	47.681	47.879	73.171	90.528	90.673	17.959	40.191	40.359	61.678	76.309	76.431
	2.0	55.975	79.384	119.88	164.93	184.09	187.95	47.183	66.915	101.05	139.02	155.17	158.43
	2.5	82.925	105.13	144.09	206.27	253.32	275.85	69.899	88.618	121.46	173.87	213.53	232.53
CCFF	0.2	2.8455	3.4662	4.6581	6.6611	9.9880	17.390	2.3985	2.9217	3.9264	5.6148	8.4191	14.666
	0.5	3.3783	7.1716	14.492	17.917	22.746	27.608	2.8476	6.0451	12.216	15.103	19.173	23.271
	1.0	5.4626	18.859	20.950	37.614	49.524	51.805	4.6046	15.897	17.659	31.706	41.745	43.668
	2.0	13.513	28.686	57.967	71.669	90.983	110.43	11.391	24.180	48.862	60.412	76.692	93.085
	2.5	19.597	35.148	64.339	109.07	117.08	130.50	16.519	29.627	54.233	91.935	98.686	110.00

Continued

Table 5.15 Effect of aspect ratios on non-dimensional frequencies of exponential FG plate with different combinations of boundary conditions—cont'd

		Ceramic constituent of Al/Al_2O_3 ($\delta = 0$)						Exponential Al/Al_2O_3 FG plate					
BCs	***a/b***	λ_1	λ_2	λ_3	λ_4	λ_5	λ_6	λ_1	λ_2	λ_3	λ_4	λ_5	λ_6
CFCF	0.2	0.6907	1.8975	2.2735	3.7425	4.7072	6.2334	0.5822	1.5995	1.9164	3.1547	3.9678	5.2543
	0.5	4.3539	7.0912	11.989	16.309	21.785	23.679	3.6700	5.9774	10.106	13.747	18.363	19.959
	1.0	17.502	20.852	34.708	48.248	53.275	64.776	14.753	17.576	29.256	40.669	44.907	54.602
	2.0	70.230	73.798	87.423	114.67	193.72	199.20	59.199	62.206	73.692	96.659	163.29	167.92
	2.5	109.82	113.39	127.02	153.44	258.69	302.94	92.568	95.578	107.07	129.34	218.07	255.36
CFSF	0.2	0.4710	1.5277	2.1591	3.2077	4.4666	5.8671	0.3970	1.2878	1.8200	2.7038	3.7650	4.9456
	0.5	2.9711	6.1647	9.6692	14.413	20.353	21.363	2.5044	5.1964	8.1504	12.149	17.156	18.007
	1.0	11.984	16.234	31.666	39.008	44.629	63.238	10.102	13.684	26.693	32.881	37.619	53.305
	2.0	48.238	52.989	69.830	102.23	156.92	162.95	40.661	44.667	58.862	86.175	132.27	137.36
	2.5	75.483	80.288	97.580	130.38	242.74	245.64	63.627	67.677	82.253	109.89	204.61	207.05
CFFF	0.2	0.1080	0.6734	1.0726	1.8875	3.2859	3.9726	0.0910	0.5676	0.9041	1.5911	2.7698	3.3486
	0.5	0.6808	2.9177	4.2303	9.5320	11.888	18.570	0.5738	2.4594	3.5658	8.0348	10.021	15.653
	1.0	2.7457	6.7112	16.794	21.717	24.526	43.705	2.3144	5.6570	14.156	18.306	20.674	36.840
	2.0	11.041	16.902	32.543	62.634	68.992	78.866	9.3065	14.247	27.431	52.796	58.156	66.478
	2.5	17.267	23.533	40.323	70.959	107.66	117.13	14.555	19.836	33.989	59.813	90.747	98.732
SCSC	0.2	17.792	18.331	19.311	20.853	27.891	37.204	14.997	15.452	16.278	17.578	23.510	31.360
	0.5	18.755	22.799	30.989	44.215	50.034	54.632	15.809	19.218	26.122	37.269	42.175	46.051
	1.0	22.799	43.112	54.596	74.580	81.679	101.84	19.218	36.340	46.021	62.866	68.849	85.839
	2.0	43.111	74.487	122.04	134.16	163.28	185.60	36.339	62.787	102.87	113.09	137.63	156.45
	2.5	59.727	90.391	137.79	201.39	203.69	232.24	50.345	76.193	116.15	169.76	171.69	195.76

SCSS	0.2	12.376	13.095	14.362	16.657	24.523	36.826		10.432	11.038	12.106	14.041	20.671	31.041
	0.5	13.649	18.623	27.835	41.031	42.956	46.215		11.505	15.698	23.463	34.586	36.209	38.956
	1.0	18.622	40.705	46.188	67.919	80.174	89.286		15.697	34.311	38.933	57.251	67.581	75.261
	2.0	40.694	67.836	111.02	133.14	159.35	177.29		34.302	57.181	93.585	112.23	134.32	149.44
	2.5	57.835	84.600	127.67	193.64	202.97	229.04		48.750	71.312	107.62	163.22	171.09	193.06
SCSF	0.2	3.0361	3.8593	5.3156	7.6423	17.728	17.783		2.5592	3.2531	4.4806	6.4419	14.943	14.989
	0.5	4.4919	9.9930	19.644	19.791	26.067	34.827		3.7863	8.4233	16.559	16.683	21.973	29.356
	1.0	9.9915	26.039	32.851	49.818	57.032	72.870		8.4221	21.949	27.691	41.993	48.074	61.424
	2.0	32.842	49.627	81.274	125.56	133.73	143.24		27.683	41.832	68.508	105.84	112.72	120.74
	2.5	50.152	66.961	98.972	151.77	195.33	213.48		42.275	56.444	83.426	127.93	164.65	179.95
SSSF	0.2	1.0591	2.3691	4.1472	6.5860	12.592	13.936		0.8928	1.9970	3.4958	5.5515	10.614	11.747
	0.5	3.1766	9.2034	14.822	19.220	21.939	33.191		2.6776	7.7578	12.494	16.201	18.494	27.977
	1.0	9.2018	21.859	32.462	46.747	49.189	72.507		7.7564	18.426	27.363	39.404	41.463	61.118
	2.0	32.445	46.522	74.986	123.32	125.49	141.15		27.349	39.215	63.208	103.95	105.78	118.98
	2.5	49.845	64.283	93.376	142.81	195.37	211.54		42.016	54.186	78.709	120.38	164.69	178.31
SFSF	0.2	0.2972	1.1951	2.0510	2.7607	4.2458	5.0731		0.2505	1.0074	1.7289	2.3271	3.5789	4.2763
	0.5	1.8728	5.4185	7.5860	12.800	17.621	20.974		1.5786	4.5674	6.3945	10.789	14.853	17.679
	1.0	7.5851	12.707	29.280	30.691	37.234	56.743		6.3937	10.711	24.681	25.870	31.386	47.830
	2.0	30.683	36.817	56.582	90.338	123.71	131.31		25.864	31.034	47.694	76.149	104.28	110.69
	2.5	48.076	54.351	75.298	110.35	193.62	201.62		40.524	45.814	63.470	93.017	163.21	169.95

Continued

Table 5.15 Effect of aspect ratios on non-dimensional frequencies of exponential FG plate with different combinations of boundary conditions—cont'd

		Ceramic constituent of Al/Al_2O_3 ($\delta = 0$)						Exponential Al/Al_2O_3 FG plate					
BCs	a/b	λ_1	λ_2	λ_3	λ_4	λ_5	λ_6	λ_1	λ_2	λ_3	λ_4	λ_5	λ_6
SSFF	0.2	0.5115	1.6366	3.0719	5.2221	11.623	12.331	0.4311	1.3795	2.5893	4.4018	9.7975	10.394
	0.5	1.3080	4.9964	11.671	12.865	17.624	23.622	1.1026	4.2116	9.8381	10.844	14.855	19.912
	1.0	2.6516	13.639	15.199	30.247	40.676	42.669	2.2351	11.497	12.811	25.496	34.287	35.967
	2.0	5.2321	19.985	46.685	51.461	70.494	94.488	4.4103	16.846	39.352	43.378	59.421	79.647
	2.5	6.4949	23.286	51.303	77.812	96.479	101.57	5.4747	19.628	43.245	65.589	81.325	85.612
SFFF	0.2	0.4644	1.0167	1.5312	3.1175	3.3623	5.5602	0.3915	0.8570	1.2907	2.6278	2.8341	4.6869
	0.5	2.5675	2.9229	8.4955	9.7022	17.099	18.296	2.1642	2.4638	7.1611	8.1782	14.413	15.422
	1.0	5.2326	11.739	20.185	20.709	39.009	40.869	4.4107	9.8954	17.015	17.457	32.881	34.449
	2.0	10.607	27.808	48.202	55.396	62.704	85.923	8.9411	23.440	40.630	46.695	52.855	72.427
	2.5	13.238	32.406	64.151	75.178	86.535	115.42	11.158	27.316	54.074	63.369	72.943	97.293
FFFF	0.2	0.6802	1.9284	2.0528	4.2735	6.7357	8.8357	0.5733	1.6255	1.7304	3.6022	5.6777	7.4478
	0.5	4.2691	5.3007	11.874	12.122	17.566	20.833	3.5985	4.4681	10.009	10.218	14.806	17.560
	1.0	10.800	15.573	19.338	28.033	28.033	49.754	9.1037	13.127	16.301	23.629	23.629	41.939
	2.0	17.067	21.203	47.497	48.489	70.262	83.331	14.394	17.872	40.036	40.873	59.226	70.242
	2.5	17.076	26.246	48.536	57.468	109.12	122.72	14.386	22.123	40.912	48.441	91.977	103.44
SFSC	0.2	3.0361	3.8593	5.3156	7.6423	17.728	17.783	2.5592	3.2531	4.4806	6.4419	14.943	14.989
	0.5	4.4919	9.9930	19.644	19.791	26.067	34.827	3.7863	8.4233	16.559	16.683	21.973	29.356
	1.0	9.9915	26.039	32.851	49.818	57.032	72.870	8.4221	21.949	27.691	41.993	48.074	61.424
	2.0	32.842	49.627	81.274	125.56	133.73	143.24	27.683	41.832	68.508	105.84	112.72	120.74
	2.5	50.152	66.961	98.972	151.77	195.33	213.48	42.275	56.444	83.426	127.93	164.65	179.95

SCCC	0.2	17.807	18.393	19.436	21.119	23.698	42.702		15.010	15.504	16.383	17.802	19.976	35.995
	0.5	19.013	23.824	32.904	48.085	50.204	55.267		16.027	20.082	27.736	40.532	42.318	46.586
	1.0	25.065	49.881	55.979	79.406	91.779	102.86		21.128	42.046	47.187	66.933	77.364	86.704
	2.0	57.802	85.229	130.09	165.83	191.19	193.23		48.723	71.842	109.66	139.78	161.16	162.88
	2.5	84.302	109.95	153.12	214.99	254.01	278.29		71.060	92.682	129.07	181.22	214.11	234.58
SCCS	0.2	12.397	13.178	14.533	16.626	21.072	39.639		10.449	11.108	12.250	14.015	17.762	33.413
	0.5	13.994	19.846	29.969	41.232	46.022	46.987		11.796	16.729	25.262	34.756	38.793	39.607
	1.0	21.306	47.681	47.879	73.171	90.528	90.673		17.959	40.191	40.359	61.678	76.309	76.431
	2.0	55.975	79.384	119.88	164.93	184.09	187.95		47.183	66.915	101.05	139.02	155.17	158.43
	2.5	82.925	105.13	144.09	206.27	253.32	275.85		69.899	88.618	121.46	173.87	213.53	232.53
CFFS	0.2	0.5663	1.8037	3.3425	5.5653	8.7082	12.219		0.4774	1.5204	2.8175	4.6911	7.3403	10.299
	0.5	1.6770	6.1013	12.634	14.008	18.423	26.835		1.4136	5.1429	10.649	11.808	15.529	22.619
	1.0	4.2211	15.029	19.447	34.097	42.151	50.362		3.5581	12.668	16.393	28.741	35.529	42.452
	2.0	12.718	24.798	50.541	71.198	88.088	98.969		10.720	20.903	42.602	60.015	74.252	83.424
	2.5	18.954	31.535	57.343	102.98	111.54	128.70		15.977	26.582	48.336	86.807	94.023	108.48
CFSC	0.2	3.0831	4.0285	5.6171	8.1424	12.205	17.755		2.5988	3.3957	4.7348	6.8634	10.288	14.966
	0.5	5.1769	11.771	19.971	22.368	27.148	38.658		4.3637	9.9223	16.834	18.855	22.884	32.586
	1.0	13.825	28.377	40.842	56.086	58.659	83.709		11.654	23.920	34.427	47.276	49.445	70.561
	2.0	49.831	63.482	92.001	141.51	158.42	173.31		42.004	53.511	77.550	119.28	133.54	146.09
	2.5	77.008	90.308	118.09	166.24	246.70	251.23		64.912	76.123	99.539	140.13	207.95	211.77

Continued

Table 5.15 Effect of aspect ratios on non-dimensional frequencies of exponential FG plate with different combinations of boundary conditions—cont'd

BCs	a/b	Ceramic constituent of Al/Al_2O_3 ($\delta = 0$)						Exponential Al/Al_2O_3 FG plate					
		λ_1	λ_2	λ_3	λ_4	λ_5	λ_6	λ_1	λ_2	λ_3	λ_4	λ_5	λ_6
CFSS	0.2	1.1735	2.6178	4.5139	7.2444	11.106	12.628	0.9892	2.2066	3.8049	6.1065	9.3613	10.644
	0.5	4.0568	11.091	15.229	21.874	23.168	35.412	3.4196	9.3490	12.837	18.438	19.529	29.849
	1.0	13.239	24.511	40.520	50.979	53.367	82.578	11.159	20.661	34.156	42.972	44.985	69.607
	2.0	49.559	60.967	86.374	134.43	158.36	171.75	41.775	51.391	72.807	113.31	133.48	144.78
	2.5	76.803	88.233	113.33	161.24	246.83	260.38	64.739	74.374	95.529	135.92	208.06	219.48
SSSS	0.2	8.0834	9.0161	10.628	12.941	22.455	31.402	6.8137	7.6000	8.9582	10.908	18.928	26.469
	0.5	9.7156	15.545	25.532	33.034	39.107	40.097	8.1895	13.103	21.522	27.845	32.964	33.799
	1.0	15.545	38.863	38.863	62.529	78.888	78.899	13.103	32.759	32.759	52.708	66.497	66.506
	2.0	38.862	62.180	102.13	132.13	156.43	160.39	32.758	52.414	86.087	111.38	131.86	135.19
	2.5	56.350	79.669	119.59	177.75	202.09	226.85	47.499	67.155	100.81	149.83	170.35	191.22
CCCF	0.2	3.1423	4.2277	5.9897	8.4500	13.056	17.779	2.6487	3.5636	5.0489	7.1227	11.005	14.987
	0.5	6.1328	13.828	20.362	25.483	28.392	40.584	5.1695	11.656	17.164	21.480	23.933	34.209
	1.0	18.869	31.516	49.843	60.449	63.544	92.815	15.905	26.566	42.014	50.954	53.563	78.236
	2.0	71.395	82.038	106.40	151.94	195.08	208.09	60.181	69.152	89.690	128.08	164.44	175.40
	2.5	110.93	121.09	143.87	186.19	256.03	304.19	93.509	102.07	121.27	156.95	215.82	256.41
CSCF	0.2	1.3110	2.9023	4.9560	7.5924	12.261	12.653	1.1050	2.4464	4.1776	6.3998	10.335	10.666
	0.5	5.2078	13.245	15.717	24.558	25.039	37.512	4.3898	11.165	13.249	20.701	21.106	31.620
	1.0	18.437	28.032	49.582	53.044	61.064	87.741	15.541	23.629	41.794	44.712	51.473	73.959
	2.0	71.205	80.058	101.47	143.37	195.06	206.69	60.021	67.483	85.530	120.85	164.42	174.23
	2.5	110.79	119.51	139.85	179.91	304.26	315.98	93.388	100.73	117.88	151.65	256.47	266.35

Source: Reproduced from Chakraverty, S., Pradhan, K.K., Free vibration of exponential functionally graded rectangular plates in thermal environment with general boundary conditions. Aerospace Science and Technology 2014;36:132–156.

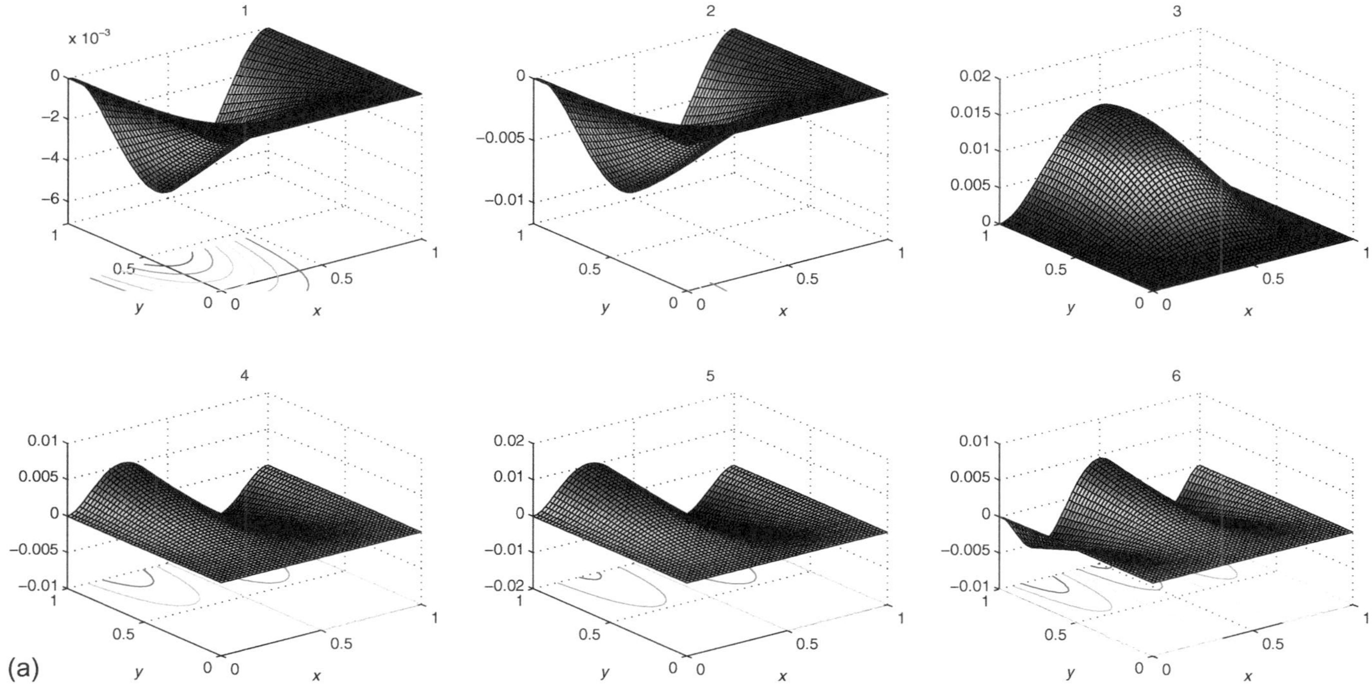

Figure 5.10 The first six mode shapes of CCCC FG rectangular plate with different $a/b = 0.2,\ 0.5$: (a) $a/b = 0.2$; *(Continued)*

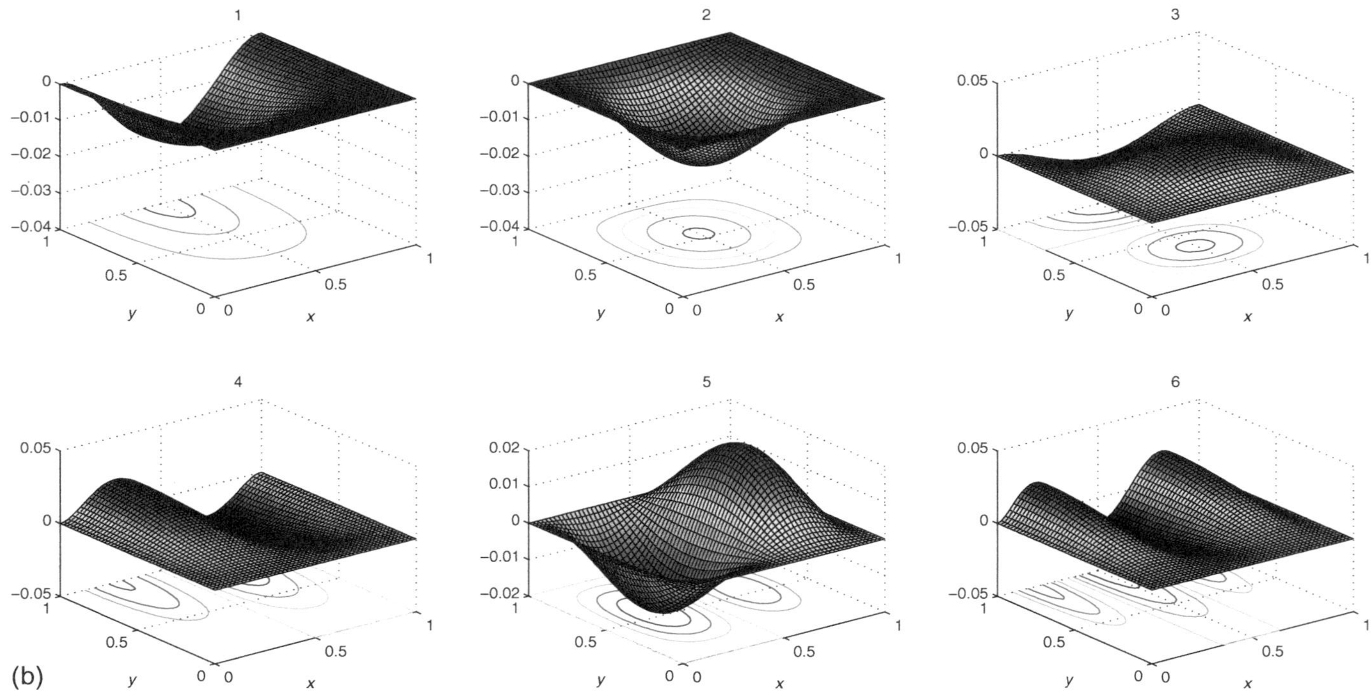

Figure 5.10, cont'd (b) $a/b = 0.5$. *(Reproduced from Chakraverty, S., Pradhan, K.K., Free vibration of exponential functionally graded rectangular plates in thermal environment with general boundary conditions. Aerospace Science and Technology 2014;36:132–156. Copyright © 2015 Elsevier Masson SAS. All rights reserved.).*

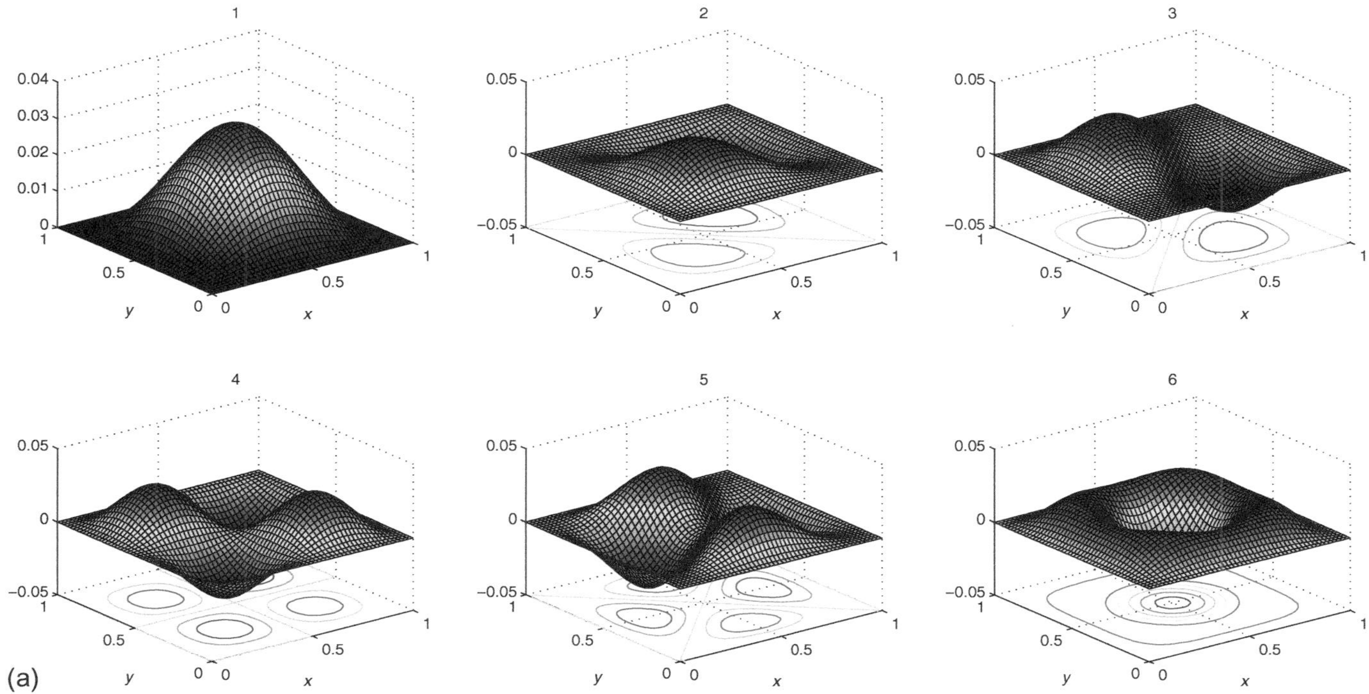

Figure 5.11 The first six mode shapes of CCCC FG rectangular plate with different $a/b = 1.0$, 2.0: (a) $a/b = 1.0$; *(Continued)*

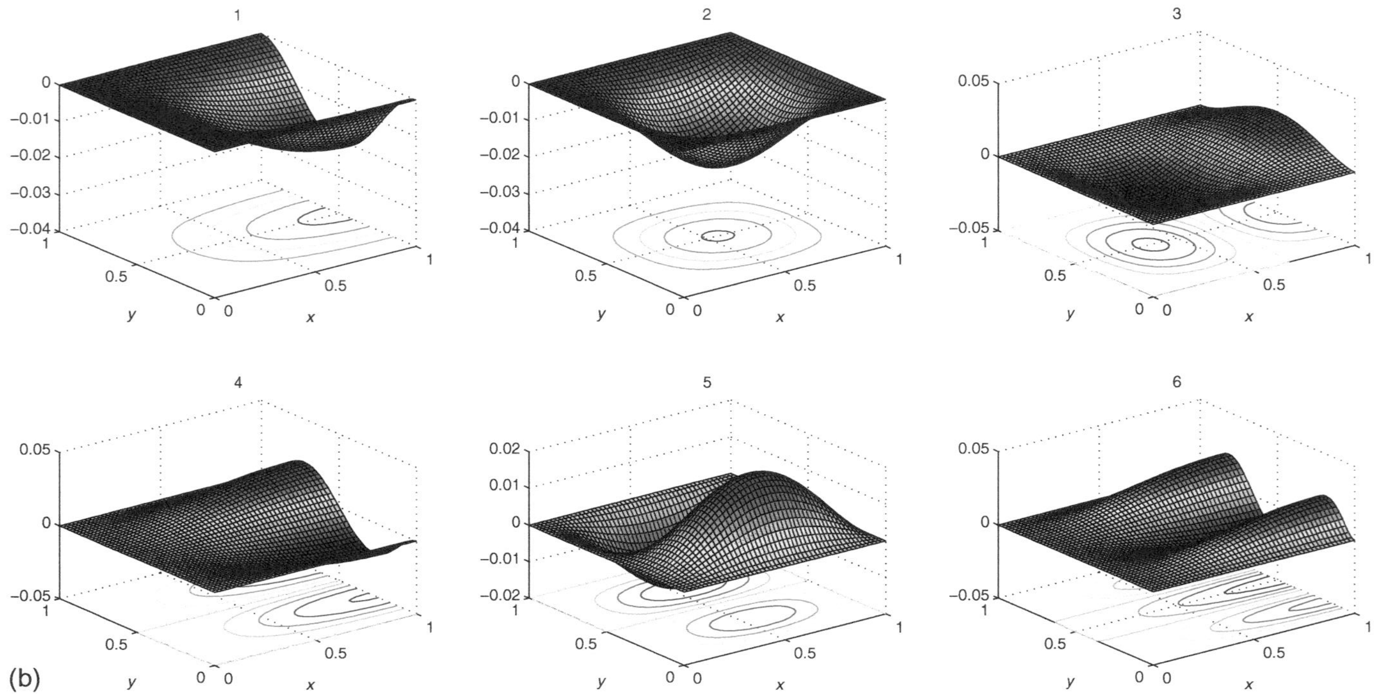

Figure 5.11, cont'd (b) $a/b = 2.0$. (*Reproduced from Chakraverty, S., Pradhan, K.K., Free vibration of exponential functionally graded rectangular plates in thermal environment with general boundary conditions. Aerospace Science and Technology 2014;36:132–156.*).

Table 5.16 Effect of E_{rat} on non-dimensional frequencies of CCCC exponential FG rectangular plates

Aspect ratio	E_{rat}	λ_1	λ_2	λ_3	λ_4	λ_5	λ_6
0.2	0.25	34.299	35.522	37.699	40.976	46.704	54.979
	0.5	27.399	28.376	30.116	32.733	37.308	43.919
	1.0	22.633	23.440	24.877	27.039	30.818	36.279
	2.0	19.374	20.065	21.295	23.146	26.381	31.056
	4.0	17.149	17.761	18.849	20.488	23.352	27.489
0.5	0.25	37.249	48.236	67.921	96.379	96.969	107.78
	0.5	29.756	38.532	54.258	76.991	77.462	86.097
	1.0	24.579	31.829	44.819	63.598	63.986	71.119
	2.0	21.041	27.246	38.366	54.441	54.774	60.879
	4.0	18.625	24.118	33.961	48.189	48.484	53.889
1.0	0.25	54.539	111.23	111.23	164.07	199.89	200.68
	0.5	43.568	88.856	88.856	131.07	159.68	160.31
	1.0	35.989	73.399	73.399	108.27	131.89	132.42
	2.0	30.807	62.831	62.831	92.677	112.91	113.36
	4.0	27.269	55.617	55.617	82.036	99.943	100.34
2.0	0.25	148.99	192.94	271.69	385.52	387.88	431.12
	0.5	119.02	154.13	217.03	307.96	309.85	344.39
	1.0	98.318	127.32	179.28	254.39	255.95	284.48
	2.0	84.162	108.99	153.46	217.76	219.09	243.52
	4.0	74.498	96.472	135.84	192.76	193.94	215.56

Source: Reproduced from Chakraverty, S., Pradhan, K.K., Free vibration of exponential functionally graded rectangular plates in thermal environment with general boundary conditions. Aerospace Science and Technology 2014;36:132–156.

Table 5.17 Effect of E_{rat} on non-dimensional frequencies of CCCS exponential FG rectangular plates

Aspect ratio	E_{rat}	λ_1	λ_2	λ_3	λ_4	λ_5	λ_6
0.2	0.25	23.901	25.535	28.345	32.519	39.242	54.426
	0.5	19.093	20.398	22.643	25.977	31.347	43.477
	1.0	15.771	16.849	18.704	21.458	25.894	35.914
	2.0	13.500	14.424	16.011	18.369	22.166	30.743
	4.0	11.950	12.767	14.173	16.259	19.621	27.213
0.5	0.25	27.808	41.003	62.588	79.778	91.977	92.960
	0.5	22.214	32.754	49.997	63.729	73.474	74.259
	1.0	18.349	27.056	41.299	52.643	60.693	61.341
	2.0	15.707	23.161	35.353	45.063	51.954	52.509
	4.0	13.904	20.502	31.294	39.889	45.989	46.480

Continued

Table 5.17 Effect of E_{rat} on non-dimensional frequencies of CCCS exponential FG rectangular plates—cont'd

Aspect ratio	E_{rat}	λ_1	λ_2	λ_3	λ_4	λ_5	λ_6
1.0	0.25	48.234	95.989	107.72	152.80	176.62	197.94
	0.5	38.531	76.679	86.053	122.06	141.09	158.12
	1.0	31.828	63.339	71.084	100.83	116.54	130.61
	2.0	27.245	54.220	60.849	86.313	99.763	111.81
	4.0	24.117	47.995	53.862	76.402	88.308	98.969
2.0	0.25	146.35	183.39	253.28	370.13	386.44	425.41
	0.5	116.91	146.49	202.33	295.67	308.69	339.83
	1.0	96.574	121.01	167.13	244.24	254.99	280.71
	2.0	82.669	103.59	143.07	209.07	218.28	240.29
	4.0	73.177	91.693	126.64	185.06	193.22	212.71

Source: Reproduced from Chakraverty, S., Pradhan, K.K., Free vibration of exponential functionally graded rectangular plates in thermal environment with general boundary conditions. Aerospace Science and Technology 2014;36:132–156.

Table 5.18 Effect of E_{rat} on non-dimensional frequencies of CCSS exponential FG rectangular plates

Aspect ratio	E_{rat}	λ_1	λ_2	λ_3	λ_4	λ_5	λ_6
0.2	0.25	23.856	25.359	27.967	31.994	40.550	76.281
	0.5	19.057	20.258	22.341	25.558	32.393	60.935
	1.0	15.742	16.734	18.454	21.112	26.758	50.335
	2.0	13.475	14.325	15.797	18.072	22.905	43.088
	4.0	11.928	12.679	13.983	15.997	20.275	38.140
0.5	0.25	26.929	38.191	57.672	79.345	88.562	90.421
	0.5	21.512	30.508	46.070	63.383	70.746	72.231
	1.0	17.769	25.201	38.056	52.357	58.439	59.665
	2.0	15.211	21.572	32.576	44.819	50.025	51.075
	4.0	13.465	19.095	28.836	39.673	44.281	45.210
1.0	0.25	41.001	91.755	92.137	140.81	174.21	174.49
	0.5	32.753	73.296	73.602	112.48	139.16	139.39
	1.0	27.055	60.546	60.798	92.914	114.95	115.14
	2.0	23.159	51.828	52.044	79.536	98.403	98.560
	4.0	20.501	45.877	46.068	70.403	87.104	87.243
2.0	0.25	107.72	152.76	230.69	317.38	354.25	361.68
	0.5	86.047	122.03	184.28	253.53	282.98	288.92
	1.0	71.079	100.80	152.22	209.43	233.76	238.66
	2.0	60.845	86.289	130.31	179.27	200.10	204.29
	4.0	53.858	76.381	115.34	158.69	177.12	180.84

Source: Reproduced from Chakraverty, S., Pradhan, K.K., Free vibration of exponential functionally graded rectangular plates in thermal environment with general boundary conditions. Aerospace Science and Technology 2014;36:132–156.

Table 5.19 Effect of E_{rat} on non-dimensional frequencies of CCCF exponential FG rectangular plates

Aspect ratio	E_{rat}	λ_1	λ_2	λ_3	λ_4	λ_5	λ_6
0.2	0.25	6.0469	8.1356	11.526	16.261	25.124	34.214
	0.5	4.8305	6.4989	9.2075	12.989	20.070	27.331
	1.0	3.9902	5.3684	7.6058	10.729	16.579	22.576
	2.0	3.4157	4.5955	6.5107	9.1850	14.192	19.326
	4.0	3.0235	4.0678	5.7631	8.1304	12.562	17.107
0.5	0.25	11.802	26.609	39.183	49.038	54.637	78.099
	0.5	9.4276	21.257	31.301	39.173	43.646	62.387
	1.0	7.7876	17.559	25.856	32.359	36.053	51.535
	2.0	6.6663	15.031	22.133	27.699	30.862	44.115
	4.0	5.9009	13.305	19.592	24.519	27.319	39.049
1.0	0.25	36.311	60.648	95.916	116.32	122.28	178.61
	0.5	29.007	48.447	76.620	92.924	97.682	142.68
	1.0	23.961	40.019	63.291	76.759	80.689	117.86
	2.0	20.511	34.257	54.179	65.707	69.072	100.89
	4.0	18.156	30.324	47.958	58.162	61.141	89.304
2.0	0.25	137.39	157.87	204.76	292.39	375.40	400.43
	0.5	109.75	126.11	163.57	233.57	299.88	319.86
	1.0	90.659	104.17	135.11	192.94	247.72	264.23
	2.0	77.606	89.174	115.66	165.16	212.05	226.19
	4.0	68.695	78.935	102.38	146.19	187.70	200.22

Source: Reproduced from Chakraverty, S., Pradhan, K.K., Free vibration of exponential functionally graded rectangular plates in thermal environment with general boundary conditions. Aerospace Science and Technology 2014;36:132–156.

Table 5.20 Effect of E_{rat} on non-dimensional frequencies of SSSS exponential FG rectangular plates

Aspect ratio	E_{rat}	λ_1	λ_2	λ_3	λ_4	λ_5	λ_6
0.2	0.25	15.555	17.350	20.451	24.903	43.211	60.428
	0.5	12.426	13.859	16.337	19.893	34.519	48.271
	1.0	10.264	11.449	13.495	16.433	28.514	39.874
	2.0	8.7865	9.8004	11.552	14.067	24.408	34.133
	4.0	7.7776	8.6751	10.226	12.451	21.606	30.214
0.5	0.25	18.696	29.914	49.133	63.569	75.256	77.160
	0.5	14.935	23.896	39.249	50.781	60.117	61.638
	1.0	12.337	19.739	32.421	41.947	49.659	50.916
	2.0	10.561	16.897	27.753	35.907	42.509	43.585
	4.0	9.3481	14.957	24.566	31.784	37.628	38.580

Continued

Table 5.20 Effect of E_{rat} on non-dimensional frequencies of SSSS exponential FG rectangular plates—cont'd

Aspect ratio	E_{rat}	λ_1	λ_2	λ_3	λ_4	λ_5	λ_6
1.0	0.25	29.914	74.786	74.786	120.33	151.81	151.83
	0.5	23.896	59.742	59.742	96.122	121.27	121.29
	1.0	19.739	49.349	49.349	79.401	100.17	100.19
	2.0	16.897	42.244	42.244	67.969	85.749	85.762
	4.0	14.957	37.393	37.393	60.164	75.904	75.915
2.0	0.25	74.785	119.66	196.53	254.27	301.02	308.64
	0.5	59.740	95.586	156.99	203.12	240.47	246.55
	1.0	49.348	78.958	129.68	167.79	198.63	203.66
	2.0	42.243	67.589	111.01	143.63	170.04	174.34
	4.0	37.392	59.829	98.265	127.14	150.51	154.32

Source: Reproduced from Chakraverty, S., Pradhan, K.K., Free vibration of exponential functionally graded rectangular plates in thermal environment with general boundary conditions. Aerospace Science and Technology 2014;36:132–156.

Table 5.21 Effect of E_{rat} on non-dimensional frequencies of FFFF exponential FG rectangular plates

Aspect ratio	E_{rat}	λ_1	λ_2	λ_3	λ_4	λ_5	λ_6
0.2	0.25	1.3089	3.7109	3.9503	8.2237	12.962	17.003
	0.5	1.0456	2.9643	3.1556	6.5693	10.354	13.583
	1.0	0.8637	2.4487	2.6067	5.4265	8.5531	11.219
	2.0	0.7393	2.0961	2.2314	4.6452	7.3216	9.6043
	4.0	0.6544	1.8554	1.9751	4.1118	6.4809	8.5015
0.5	0.25	8.2152	10.201	22.850	23.328	33.802	40.089
	0.5	6.5626	8.1484	18.254	18.635	27.002	32.025
	1.0	5.4209	6.7309	15.078	15.393	22.305	26.454
	2.0	4.6404	5.7618	12.907	13.177	19.094	22.645
	4.0	4.1076	5.1002	11.425	11.664	16.901	20.045
1.0	0.25	20.783	29.967	37.214	53.946	53.946	95.745
	0.5	16.602	23.939	29.728	43.093	43.093	76.484
	1.0	13.714	19.774	24.556	35.597	35.597	63.179
	2.0	11.739	16.927	21.021	30.472	30.472	54.082
	4.0	10.392	14.984	18.607	26.973	26.973	47.872
2.0	0.25	32.861	40.802	91.401	93.310	135.21	160.36
	0.5	26.250	32.594	73.014	74.539	108.01	128.09
	1.0	21.684	26.924	60.312	61.572	89.219	105.81
	2.0	18.562	23.047	51.629	52.707	76.374	90.579
	4.0	16.430	20.401	45.701	46.655	67.605	80.179

Source: Reproduced from Chakraverty, S., Pradhan, K.K., Free vibration of exponential functionally graded rectangular plates in thermal environment with general boundary conditions. Aerospace Science and Technology 2014;36:132–156.

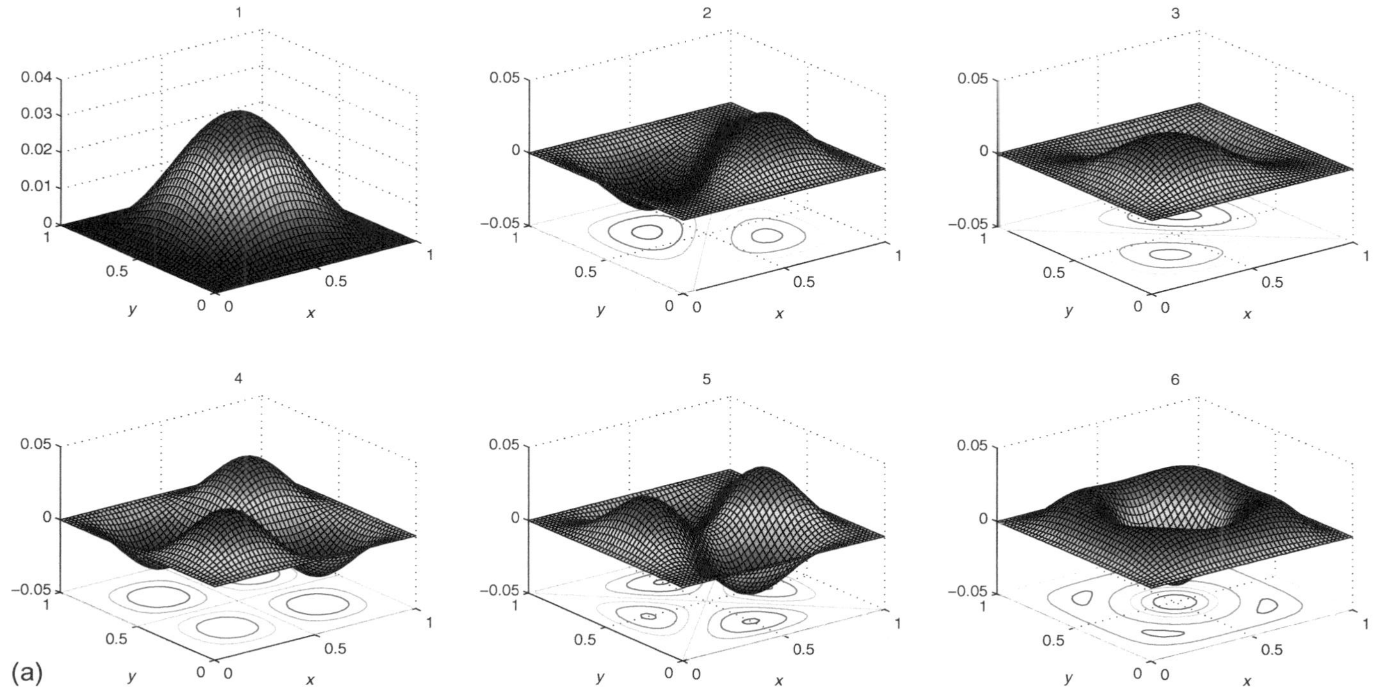

Figure 5.12 The first six mode shapes of CCCC FG rectangular plate with $a/b = 1.0$ and $E_{rat} = 0.25, 0.5$: (a) $E_{rat} = 0.25$; (*Continued*)

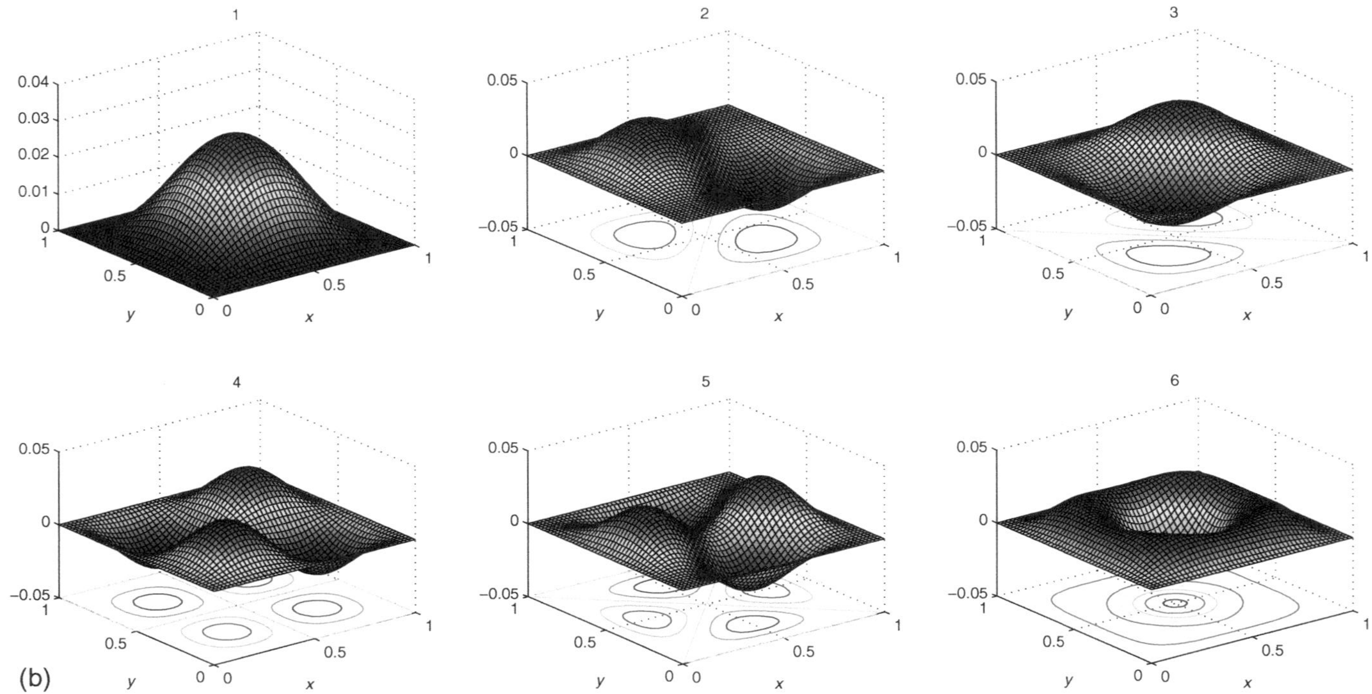

Figure 5.12, cont'd (b) $E_{rat} = 0.5$. (*Reproduced from Chakraverty, S., Pradhan, K.K., Free vibration of exponential functionally graded rectangular plates in thermal environment with general boundary conditions. Aerospace Science and Technology 2014;36:132–156. Copyright © 2015 Elsevier Masson SAS. All rights reserved.*).

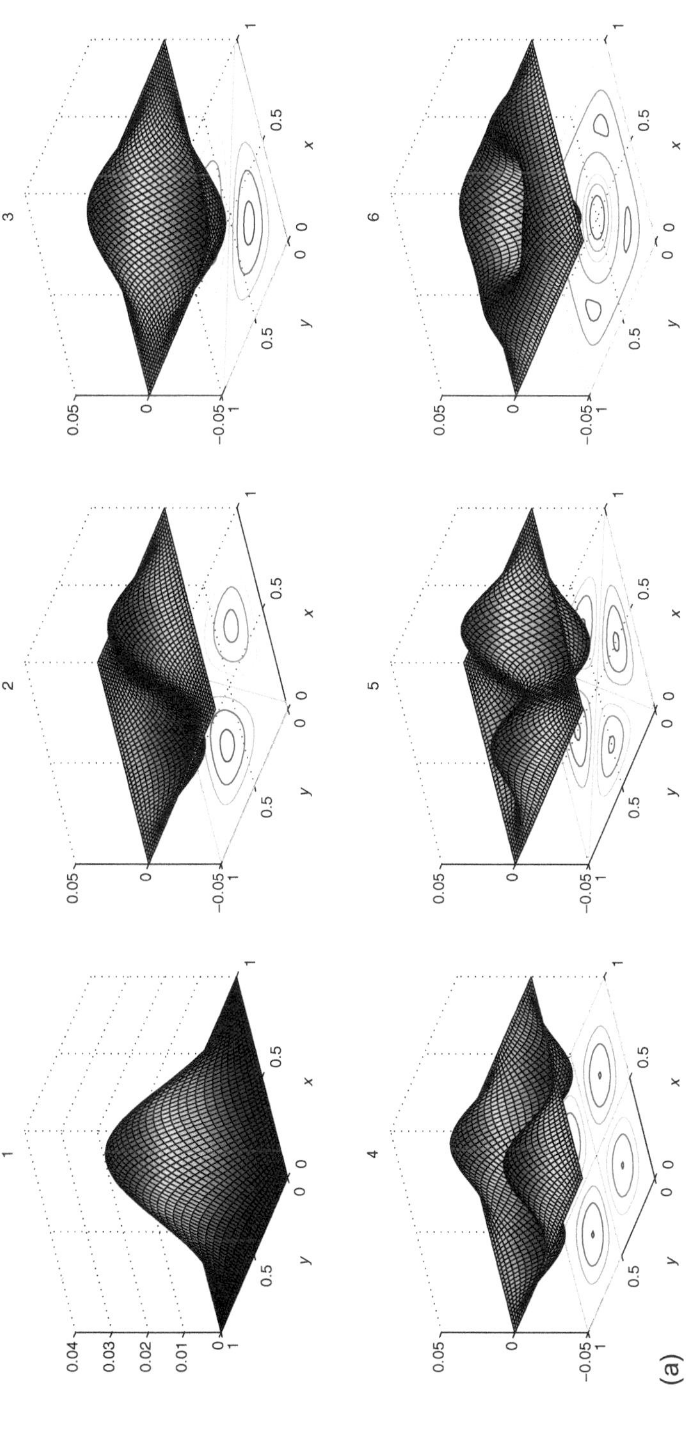

(a)

Figure 5.13 The first six mode shapes of CCCC FG rectangular plates with $a/b = 1.0$ and $E_{\mathrm{rat}} = 1.0, 2.0$: (a) $E_{\mathrm{rat}} = 1.0$; *(Continued)*

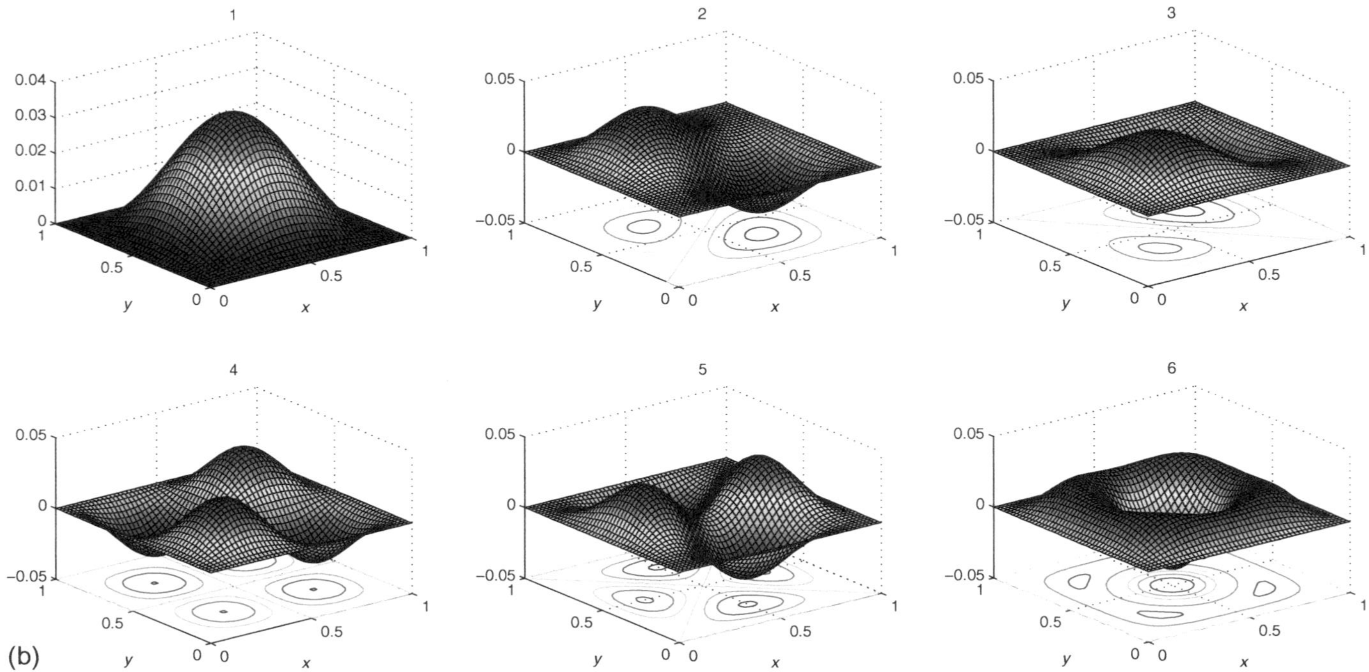

Figure 5.13, cont'd (b) $E_{rat} = 2.0$. (*Reproduced from Chakraverty, S., Pradhan, K.K., Free vibration of exponential functionally graded rectangular plates in thermal environment with general boundary conditions. Aerospace Science and Technology 2014;36:132–156.*).

Table 5.22 Effect of ρ_{rat} on non-dimensional frequencies of CCCC exponential FG rectangular plates

Aspect ratio	ρ_{rat}	λ_1	λ_2	λ_3	λ_4	λ_5	λ_6
	0.25	15.385	15.934	16.911	18.380	20.949	24.662
	0.5	18.843	19.515	20.711	22.511	25.658	30.204
0.2	1.0	22.633	23.439	24.877	27.039	30.818	36.279
	2.0	26.648	27.598	29.290	31.836	36.285	42.715
	4.0	30.771	31.868	33.821	36.761	41.899	49.323
	0.25	16.709	21.637	30.467	43.232	43.497	48.346
	0.5	20.464	26.499	37.314	52.948	53.272	59.211
0.5	1.0	24.579	31.829	44.819	63.598	63.986	71.119
	2.0	28.940	37.476	52.770	74.880	75.338	83.737
	4.0	33.417	43.274	60.934	86.464	86.993	96.691
	0.25	24.464	49.895	49.895	73.596	89.662	90.019
	0.5	29.963	61.109	61.109	90.137	109.81	110.25
1.0	1.0	35.989	73.399	73.399	108.27	131.89	132.42
	2.0	42.374	86.421	86.421	127.47	155.29	155.92
	4.0	48.929	99.789	99.789	147.19	179.32	180.04
	0.25	66.834	86.547	121.87	172.93	173.99	193.38
	0.5	81.855	105.99	149.26	211.79	213.09	236.84
2.0	1.0	98.318	127.32	179.28	254.39	255.95	284.48
	2.0	115.76	149.90	211.08	299.52	301.35	334.95
	4.0	133.67	173.09	243.74	345.86	347.97	386.76

Source: Reproduced from Chakraverty, S., Pradhan, K.K., Free vibration of exponential functionally graded rectangular plates in thermal environment with general boundary conditions. Aerospace Science and Technology 2014;36:132–156.

$E_{rat} = 1$. In Table 5.22, frequencies of FG plates with the variation of ρ_{rat}, viz. $\rho_{rat} = 0.25$, 0.5, 1.0, and 2.0, are evaluated with a fixed aspect ratio. In a similar fashion, effect of ρ_{rat} on these frequencies are incorporated in Tables 5.23–5.27 with various edge conditions, viz. CCCS, CCSS, CCCF, SSSS, and FFFF. One may easily conclude that frequencies are increasing with and increase in ρ_{rat} irrespective of the aspect ratio and BC assumed. The authenticity of this argument can also be observed from the expression of non-dimensional frequency, which is directly proportional to the mass density of the ceramic constituent of FG rectangular plates.

Accordingly, the first six mode shapes of CCCC FG square plates ($a/b = 1$) are given in Figs. 5.14 and 5.15 with different ρ_{rat}, viz. $\rho_{rat} =$ 0.25, 0.5, 1.0, and 2.0. In a similar way, 3D mode shapes of exponential FG plate subjected to various BCs with different ρ_{rat} may also be depicted.

Table 5.23 Effect of ρ_{rat} on non-dimensional frequencies of CCCS exponential FG rectangular plates

Aspect ratio	ρ_{rat}	λ_1	λ_2	λ_3	λ_4	λ_5	λ_6
0.2	0.25	10.721	11.454	12.715	14.587	17.602	24.413
	0.5	13.130	14.028	15.572	17.865	21.558	29.900
	1.0	15.771	16.849	18.704	21.458	25.894	35.914
	2.0	18.569	19.839	22.022	25.265	30.488	42.285
	4.0	21.442	22.908	25.429	29.174	35.205	48.827
0.5	0.25	12.474	18.392	28.075	35.785	41.257	41.698
	0.5	15.277	22.526	34.384	43.828	50.529	51.069
	1.0	18.349	27.056	41.299	52.643	60.693	61.341
	2.0	21.605	31.857	48.627	61.982	71.460	72.224
	4.0	24.947	36.785	56.149	71.570	82.515	83.397
1.0	0.25	21.636	43.057	48.321	68.542	79.224	88.788
	0.5	26.498	52.734	59.181	83.947	97.029	108.74
	1.0	31.828	63.339	71.084	100.83	116.54	130.61
	2.0	37.474	74.577	83.694	118.72	137.22	153.79
	4.0	43.272	86.114	96.642	137.08	158.45	177.58
2.0	0.25	65.649	82.259	113.61	166.03	173.34	190.82
	0.5	80.403	100.75	139.14	203.34	212.29	233.71
	1.0	96.574	121.01	167.13	244.24	254.99	280.71
	2.0	113.71	142.48	196.78	287.57	300.24	330.52
	4.0	131.29	164.52	227.22	332.05	346.68	381.65

Source: Reproduced from Chakraverty, S., Pradhan, K.K., Free vibration of exponential functionally graded rectangular plates in thermal environment with general boundary conditions. Aerospace Science and Technology 2014;36:132–156.

Table 5.24 Effect of ρ_{rat} on non-dimensional frequencies of CCSS exponential FG rectangular plates

Aspect ratio	ρ_{rat}	λ_1	λ_2	λ_3	λ_4	λ_5	λ_6
0.2	0.25	10.701	11.375	12.545	14.351	18.189	34.217
	0.5	13.106	13.932	15.364	17.577	22.277	41.907
	1.0	15.742	16.734	18.454	21.112	26.758	50.335
	2.0	18.534	19.703	21.728	24.857	31.505	59.265
	4.0	21.402	22.751	25.089	28.703	36.379	68.433
0.5	0.25	12.079	17.131	25.869	35.591	39.726	40.559
	0.5	14.794	20.981	31.683	43.590	48.654	49.675
	1.0	17.769	25.201	38.056	52.357	58.439	59.665
	2.0	20.922	29.672	44.807	61.646	68.807	70.251
	4.0	24.159	34.262	51.739	71.182	79.451	81.118

Table 5.24 Effect of ρ_{rat} on non-dimensional frequencies of CCSS exponential FG rectangular plates—cont'd

Aspect ratio	ρ_{rat}	λ_1	λ_2	λ_3	λ_4	λ_5	λ_6
	0.25	18.392	41.158	41.329	63.161	78.143	78.268
	0.5	22.525	50.408	50.618	77.356	95.705	95.859
1.0	1.0	27.055	60.546	60.798	92.914	114.95	115.14
	2.0	31.855	71.287	71.584	109.39	135.35	135.56
	4.0	36.783	82.315	82.658	126.32	156.29	156.54
	0.25	48.318	68.523	103.48	142.36	158.90	162.24
	0.5	59.177	83.924	126.73	174.36	194.62	198.69
2.0	1.0	71.079	100.80	152.22	209.43	233.76	238.66
	2.0	83.689	118.69	179.23	246.58	275.23	281.00
	4.0	96.635	137.05	206.96	284.73	317.81	324.47

Source: Reproduced from Chakraverty, S., Pradhan, K.K., Free vibration of exponential functionally graded rectangular plates in thermal environment with general boundary conditions. Aerospace Science and Technology 2014;36:132–156.

Table 5.25 Effect of ρ_{rat} on non-dimensional frequencies of CCCF exponential FG rectangular plates

Aspect ratio	ρ_{rat}	λ_1	λ_2	λ_3	λ_4	λ_5	λ_6
	0.25	2.7124	3.6493	5.1703	7.2939	11.269	15.347
	0.5	3.3220	4.4695	6.3322	8.9333	13.803	18.796
0.2	1.0	3.9902	5.3684	7.6058	10.729	16.579	22.576
	2.0	4.6981	6.3208	8.9551	12.634	19.519	26.582
	4.0	5.4249	7.2987	10.341	14.588	22.539	30.694
	0.25	5.2938	11.936	17.576	21.997	24.508	35.032
	0.5	6.4836	14.619	21.526	26.940	30.016	42.905
0.5	1.0	7.7876	17.559	25.856	32.359	36.053	51.535
	2.0	9.1691	20.674	30.443	38.099	42.449	60.677
	4.0	10.588	23.872	35.152	43.993	49.016	70.064
	0.25	16.288	27.204	43.024	52.179	54.851	80.117
	0.5	19.948	33.318	52.694	63.906	67.178	98.123
1.0	1.0	23.961	40.019	63.291	76.759	80.689	117.86
	2.0	28.211	47.119	74.519	90.376	95.005	138.77
	4.0	32.576	54.409	86.048	104.36	109.70	160.23
	0.25	61.628	70.814	91.847	131.16	168.39	179.62
	0.5	75.478	86.729	112.49	160.63	206.24	219.99
2.0	1.0	90.659	104.17	135.11	192.94	247.72	264.23
	2.0	106.74	122.65	159.08	227.17	291.66	311.11
	4.0	123.26	141.63	183.69	262.31	336.78	359.24

Source: Reproduced from Chakraverty, S., Pradhan, K.K., Free vibration of exponential functionally graded rectangular plates in thermal environment with general boundary conditions. Aerospace Science and Technology 2014;36:132–156.

Table 5.26 Effect of ρ_{rat} on non-dimensional frequencies of SSSS exponential FG rectangular plates

Aspect ratio	ρ_{rat}	λ_1	λ_2	λ_3	λ_4	λ_5	λ_6
0.2	0.25	6.9775	7.7827	9.1736	11.170	19.383	27.106
	0.5	8.5457	9.5318	11.235	13.681	23.739	33.197
	1.0	10.264	11.449	13.495	16.433	28.514	39.874
	2.0	12.085	13.480	15.889	19.348	33.572	46.948
	4.0	13.955	15.565	18.347	22.341	38.766	54.211
0.5	0.25	8.3864	13.418	22.039	28.514	33.757	34.611
	0.5	10.271	16.434	26.992	34.923	41.344	42.389
	1.0	12.337	19.739	32.421	41.947	49.659	50.916
	2.0	14.526	23.241	38.173	49.389	58.469	59.948
	4.0	16.773	26.837	44.078	57.029	67.514	69.223
1.0	0.25	13.418	33.546	33.546	53.975	68.095	68.105
	0.5	16.434	41.086	41.086	66.105	83.399	83.411
	1.0	19.739	49.349	49.349	79.401	100.17	100.19
	2.0	23.241	58.104	58.104	93.487	117.94	117.96
	4.0	26.837	67.093	67.093	107.95	136.19	136.21
2.0	0.25	33.546	53.674	88.156	114.06	135.03	138.45
	0.5	41.085	65.737	107.97	139.69	165.37	169.56
	1.0	49.348	78.958	129.68	167.79	198.63	203.66
	2.0	58.103	92.966	152.69	197.55	233.87	239.79
	4.0	67.091	107.35	176.31	228.12	270.06	276.89

Source: Reproduced from Chakraverty, S., Pradhan, K.K., Free vibration of exponential functionally graded rectangular plates in thermal environment with general boundary conditions. Aerospace Science and Technology 2014;36:132–156.

Table 5.27 Effect of ρ_{rat} on non-dimensional frequencies of FFFF exponential FG rectangular plates

Aspect ratio	ρ_{rat}	λ_1	λ_2	λ_3	λ_4	λ_5	λ_6
0.2	0.25	0.5871	1.6645	1.7719	3.6888	5.8142	7.6269
	0.5	0.7191	2.0386	2.1702	4.5179	7.1209	9.3409
	1.0	0.8637	2.4487	2.6067	5.4265	8.5531	11.219
	2.0	1.0169	2.8831	3.0691	6.3893	10.070	13.210
	4.0	1.1742	3.3291	3.5439	7.3777	11.628	15.254
0.5	0.25	3.6850	4.5755	10.249	10.464	15.162	17.983
	0.5	4.5132	5.6039	12.553	12.816	18.570	22.024
	1.0	5.4209	6.7309	15.078	15.393	22.305	26.454
	2.0	6.3827	7.9251	17.753	18.124	26.262	31.147
	4.0	7.3701	9.1511	20.499	20.928	30.325	35.965

Table 5.27 Effect of ρ_{rat} on non-dimensional frequencies of FFFF exponential FG rectangular plates—cont'd

Aspect ratio	ρ_{rat}	λ_1	λ_2	λ_3	λ_4	λ_5	λ_6
1.0	0.25	9.3225	13.442	16.693	24.198	24.198	42.947
	0.5	11.418	16.463	20.444	29.636	29.636	52.599
	1.0	13.714	19.774	24.556	35.597	35.597	63.179
	2.0	16.147	23.283	28.913	41.912	41.912	74.387
	4.0	18.645	26.884	33.385	48.396	48.396	85.895
2.0	0.25	14.740	18.302	40.999	41.855	60.649	71.931
	0.5	18.053	22.415	50.213	51.262	74.280	88.097
	1.0	21.684	26.924	60.312	61.572	89.219	105.81
	2.0	25.531	31.700	71.012	72.496	105.05	124.59
	4.0	29.480	36.604	81.998	83.711	121.29	143.86

Source: Reproduced from Chakraverty, S., Pradhan, K.K., Free vibration of exponential functionally graded rectangular plates in thermal environment with general boundary conditions. Aerospace Science and Technology 2014;36:132–156.

5.4 CONCLUDING REMARKS

The present study involves the evaluation of natural frequencies and mode shapes of thin FG plates with various boundary conditions based on classical plate theory. Both power and exponential gradation of FG material properties have been considered. In addition, the vibration problem associated with isotropic thick rectangular plates subjected to different edge conditions has also been investigated. The generalized eigenvalue problem for free vibration can be obtained by means of the Rayleigh-Ritz method, which may handle any combination of boundary conditions easily. Looking into the numerical formulation and results, the following facts may be concluded.

- The aspect ratios, power-law indices, and different material distributions are key factors when examining free vibration characteristics of FG rectangular and square plates.
- In the Rayleigh-Ritz method, the increase in the number of polynomials (n) plays a crucial role in the convergence of non-dimensional frequencies.
- From the tables and figures presented in this, it may be observed that natural frequencies in the case of FG plates are increasing with an increase in aspect ratios for a fixed power-law index and are decreasing with an increase in power-law exponents for a fixed aspect ratio. This

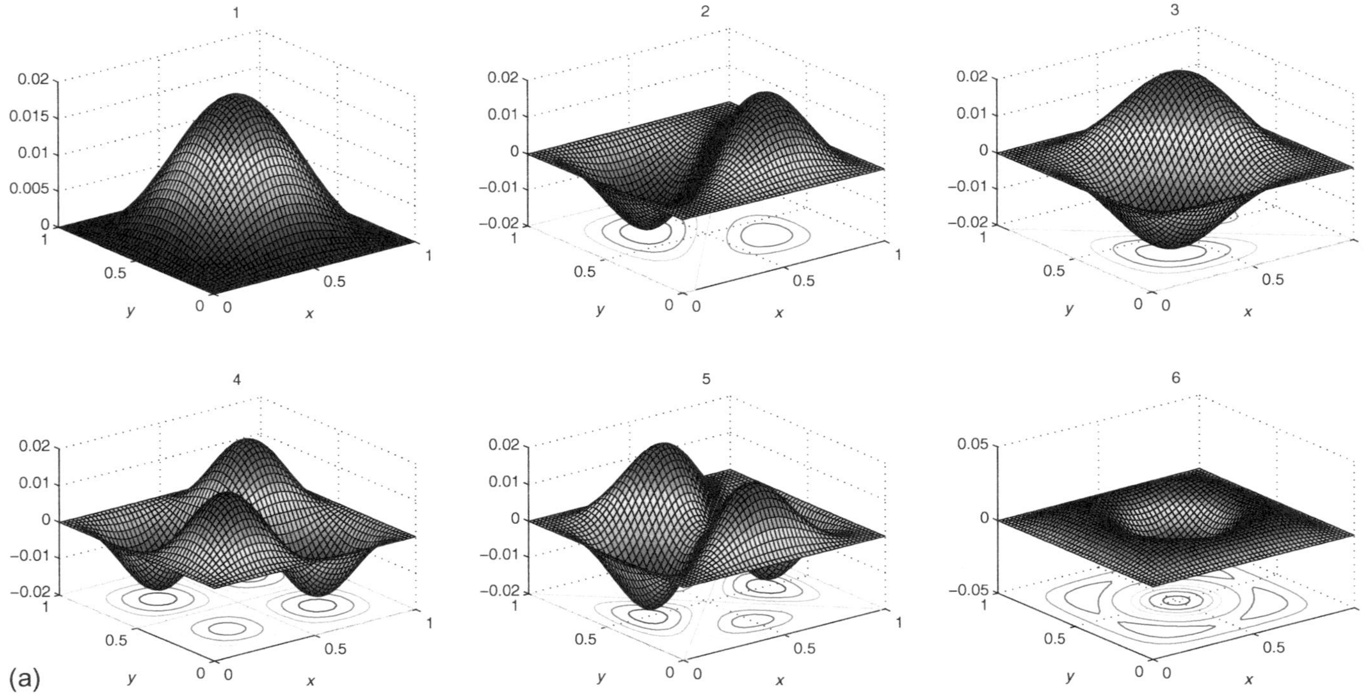

Figure 5.14 The first six mode shapes of CCCC FG rectangular plate with $a/b = 1.0$ and $\rho_{\text{rat}} = 0.25, 0.5$: (a) $\rho_{\text{rat}} = 0.25$; *(Continued)*

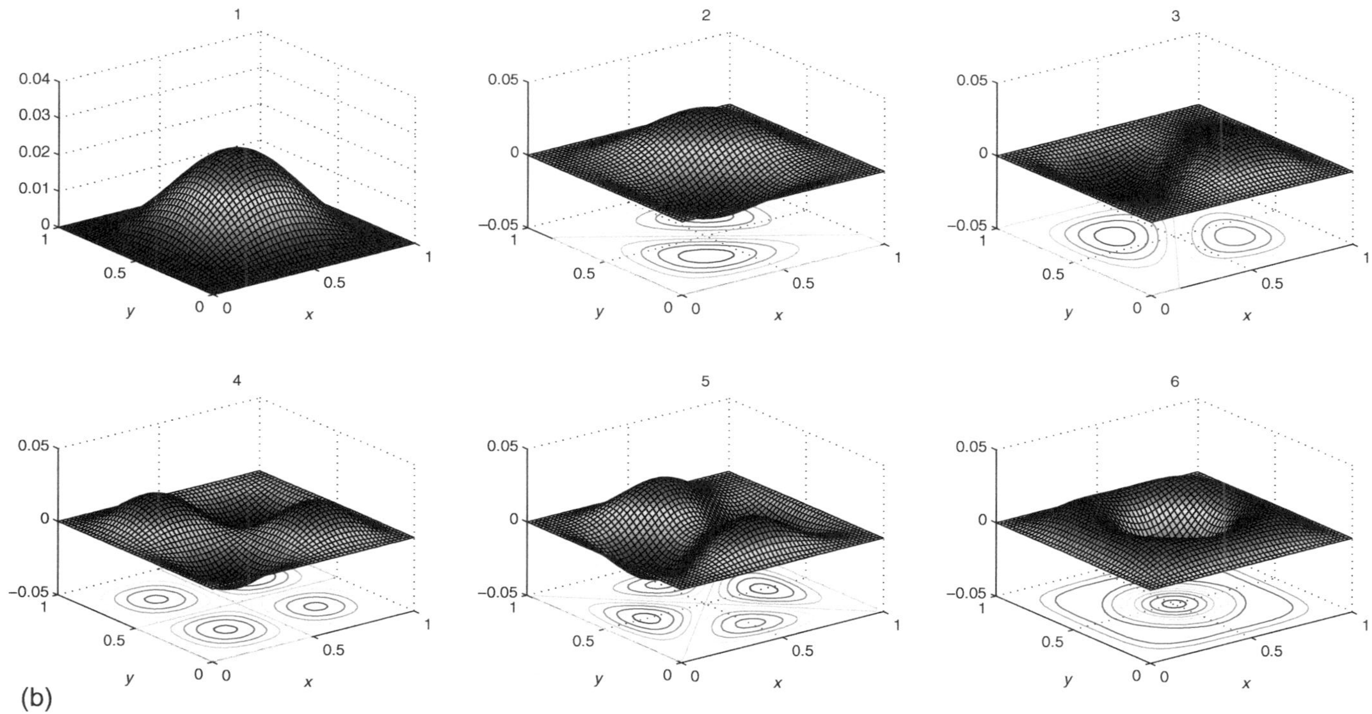

Figure 5.14, cont'd (b) $\rho_{\text{rat}} = 0.5$. (*Reproduced from Chakraverty, S., Pradhan, K.K., Free vibration of exponential functionally graded rectangular plates in thermal environment with general boundary conditions. Aerospace Science and Technology 2014;36:132–156.*).

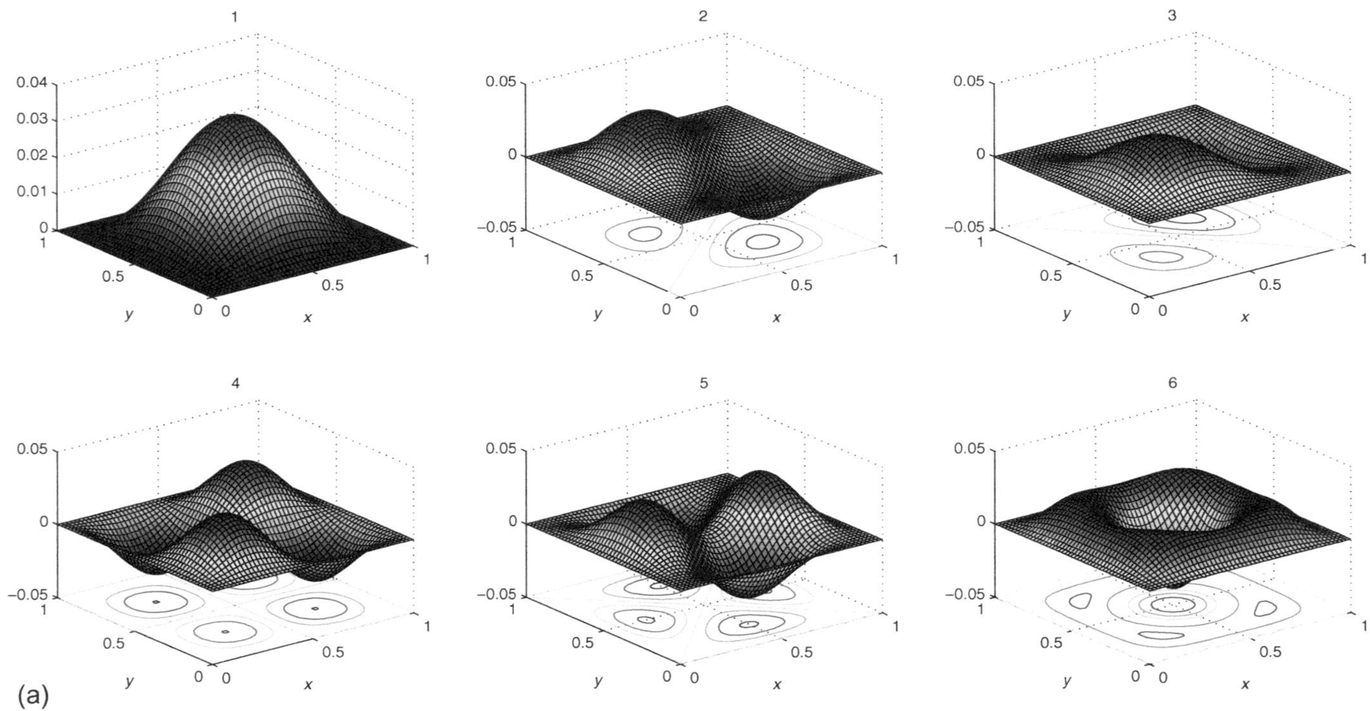

Figure 5.15 The first six mode shapes of CCCC FG rectangular plates with $a/b = 1.0$ and $\rho_{\text{rat}} = 1.0, 2.0$: (a) $\rho_{\text{rat}} = 1.0$; (*Continued*)

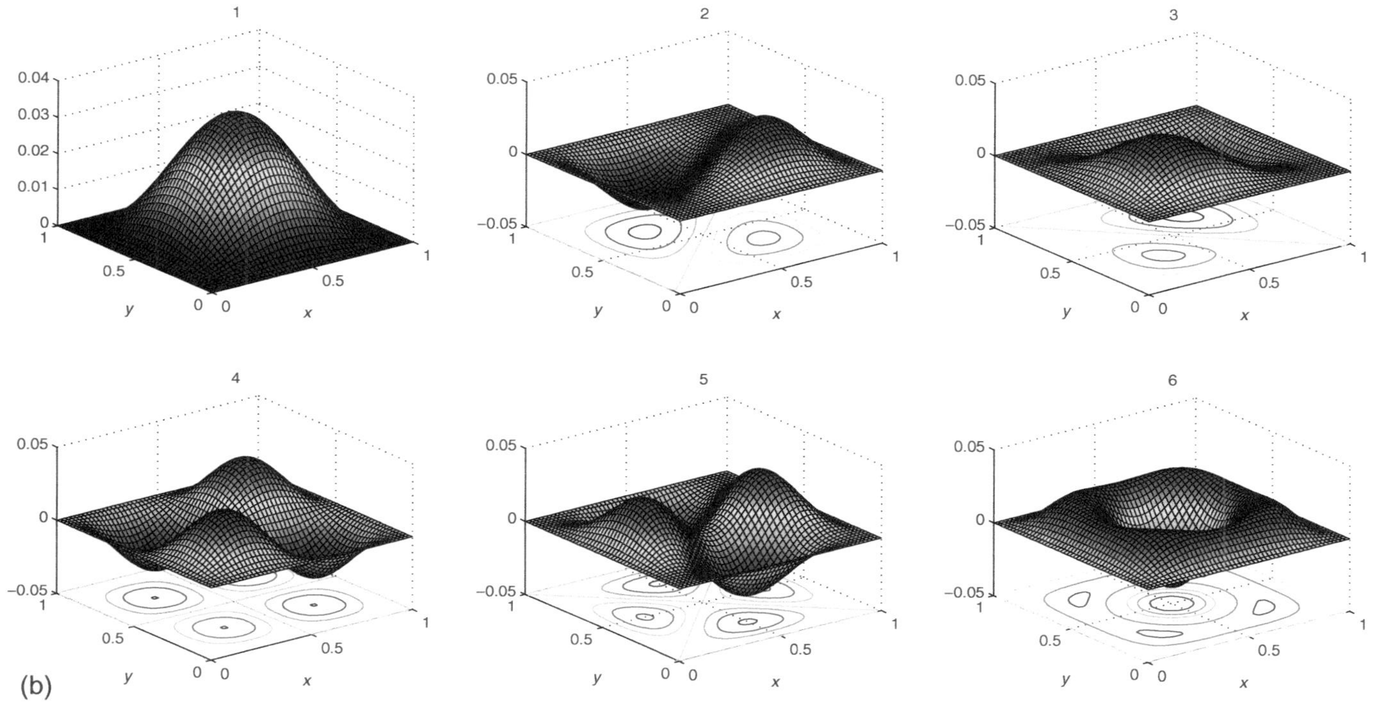

Figure 5.15, cont'd (b) $\rho_{rat} = 2.0$. (*Reproduced from Chakraverty, S., Pradhan, K.K., Free vibration of exponential functionally graded rectangular plates in thermal environment with general boundary conditions. Aerospace Science and Technology 2014;36:132–156. Copyright © 2015 Elsevier Masson SAS. All rights reserved.*).

behavior is true irrespective of the edge support, gradation criteria, and configuration of the FG plate considered.

- On the other hand, in the case of exponential gradation, the frequencies concerned are decreasing with an increase in E_{rat} and behave reverse with an increase in ρ_{rat}. The authenticity of these arguments may be found from the expression of natural frequency.
- Other shear deformation plate theories can also be extended easily following the above analysis.

CHAPTER 6

Vibration Problems of Functionally Graded Elliptic Plates

In particular, a significant number of works on vibration behavior of isotropic elliptic and circular plates may also be found in open literature, whereas detailed analyses on thin functionally graded (FG) elliptic and circular plates are very limited. A few of the major contributions on these studies have been reported here. Accordingly, Mazumdar (1971) has computed fundamental frequencies of elliptic plates subjected to both clamped and simply supported edge supports by the method of constant deflection lines. Leissa and Narita (1980) have analyzed the natural frequencies of simply supported circular plates using classical plate theory with ordinary and modified Bessel functions of the first kind. Galerkin and Bolotin's methods are used by Chen and Hwang (1988) to study dynamic stability of isotropic Mindlin circular plates subjected to periodic radial loads. The spline-finite-strip method with subparametric mapping concept is employed by Cheung and Tham (1988) in static and free vibration analysis of arbitrary shaped plates. Transverse vibration of isotropic elliptic plates subjected to different classical boundary supports can be found in Singh and Chakraverty (1991, 1992a,b). Circular and elliptic plates with variable thickness have been investigated in Singh and Chakraverty (1994b) with all three classical boundary conditions. Natural frequencies for free vibration of nonhomogeneous circular and elliptic plates using two dimensional orthogonal polynomials are studied in Chakraverty and Petyt (1997). Liew et al. (1997) used differential quadrature method and linear shear deformation Mindlin theory to analyze axisymmetric free vibration characteristics of moderately thick circular plates. Reddy et al. (1999) have used first-order shear deformation Mindlin plate theory to study axisymmetric bending and stretching of FG solid and annular circular plates. Three-dimensional vibration of circular and annular plates are analyzed in Liu and Lee (2000) using finite element method and in Zhao et al. (2003) applying the Chebyshev-Ritz method. Free vibration of solid circular plates has been studied in Wu and Liu (2001) and Wu et al. (2002) by applying

Vibration of Functionally Graded Beams and Plates
http://dx.doi.org/10.1016/B978-0-12-804228-1.00006-7

generalized differential quadrature rule. Buckling analysis is presented in Najafizadeh and Eslami (2002) for radially loaded FG solid circular plate subject to either clamped or simply supported edge conditions. Ma and Wang (2003) investigated axisymmetric large deflection bending and thermal postbuckling behavior of FG circular plate under mechanical, thermal, and combined thermomechanical loadings based on classical nonlinear von Karman plate theory. An inverse problem of a FG elliptic plate is developed by Hsieh and Lee (2006) with large deflection and distributed boundary under uniform load. Prakash and Ganpathi (2006) have found free vibration characteristics and thermoelastic stability of FG circular plates using finite element procedure. Chakraverty et al. (2007) have provided vibration behavior of plates by using the effects of nonhomogeneity.

Based on the above survey, free vibration of thin FG elliptic plates subjected to various classical boundary conditions has been discussed in this chapter by using the Rayleigh-Ritz method (as stated in Chapter 3). The stress-strain relations are considered based on classical plate theory. The power-law form along thickness direction is assumed for gradation of material properties of FG elliptic plate constituents. Trial functions denoting the displacement components are expressed as simple algebraic polynomials, which can handle any classical edge support with ease. The tests of convergence of natural frequencies are summarized along with the validation in special cases. Consequently, the effect of geometric configurations and gradation of constituent volume fractions on the natural frequencies can be estimated. Moreover, 3D mode shapes for circular and elliptic FG plates with various boundary conditions at the edges are also demonstrated.

6.1 NUMERICAL MODELING

The mathematical modeling for free vibration of FG thin elliptic plates follows the systematic procedures mentioned in Section 3.3.2 with power-law variation of material properties as stated in Eq. (2.1). On the contrary, the admissible function of FG elliptic plate is also given in the modeling part concerned. Based on the numerical formulation, the natural frequencies for free vibration of FG elliptic plates may now be evaluated after getting the generalized eigenvalue problem of the form Eq. (3.26). 3D mode shapes are also presented to show the deflected shapes of elliptic plates with respect to a few boundary supports. The detailed results have already been computed in Pradhan and Chakraverty (2014b) and are incorporated in further sections.

6.2 CONVERGENCE AND COMPARISON STUDIES

In this part, convergence studies for non-dimensional frequencies of isotropic elliptic (or circular) plates (assuming $k = 0$ in the case of FG plates) are reported with an increase in the number of polynomials involved in the displacement component. The material properties of FG constituents are considered in Table 6.1.

The non-dimensional frequencies of elliptic FG plate may be written as

$$\lambda = \omega a^2 \sqrt{\frac{\rho_c h}{D_c}} \tag{6.1}$$

where D_c (flexural rigidity) $= \frac{E_c h^3}{12(1-\nu^2)}$ of the FG plate due to the deformation effect.

In Tables 6.2 and 6.3, convergences of the first six non-dimensional frequencies of isotropic circular and elliptic plates are incorporated. Isotropic

Table 6.1 Material properties of the FG elliptic plate constituents

Properties	**Unit**	**Aluminum (Al)**	**Alumina (Al_2O_3)**
E	GPa	70	380
ρ	kg/m^3	2700	3800
ν	-	0.3	0.3

From Pradhan, K.K., Chakraverty, S., 2014. Free vibration of functionally graded thin elliptic plates with various edge supports. Structural Engineering and Mechanics 53 (2), 337–354.

Table 6.2 Convergence of non-dimensional frequencies for isotropic circular plate

BCs	**Sources**	λ_1	λ_2	λ_3	λ_4	λ_5	λ_6
C	10 × 10	10.217	21.275	36.661	43.058	54.650	69.202
	13 × 13	10.216	21.275	35.609	41.210	54.650	69.202
	16 × 16	10.216	21.266	34.941	39.921	52.479	64.682
	19 × 19	10.216	21.263	34.941	39.921	51.914	62.439
	20 × 20	10.216	21.261	34.941	39.921	51.209	61.407
S	10 × 10	4.941	13.987	35.665	46.706	59.195	88.161
	13 × 13	4.938	13.987	30.391	39.456	59.195	88.161
	16 × 16	4.935	13.941	25.986	30.503	46.102	59.195
	19 × 19	4.935	13.915	25.986	30.503	42.362	45.976
	20 × 20	4.935	13.899	25.986	30.503	40.915	42.362

From Pradhan, K.K., Chakraverty, S., 2014. Free vibration of functionally graded thin elliptic plates with various edge supports. Structural Engineering and Mechanics 53 (2), 337–354.

Table 6.3 Convergence of non-dimensional frequencies for isotropic elliptic plate with $a/b = 2$

BCs	Sources	λ_1	λ_2	λ_3	λ_4	λ_5	λ_6
C	10 × 10	27.395	39.594	61.455	70.023	88.595	95.687
	13 × 13	27.378	39.594	56.329	70.023	88.595	89.812
	16 × 16	27.378	39.503	56.328	70.023	78.029	88.665
	19 × 19	27.378	39.500	56.328	69.884	78.025	88.665
	20 × 20	27.378	39.499	56.328	69.884	78.020	88.665
S	10 × 10	13.258	23.910	46.747	55.136	91.258	93.936
	13 × 13	13.218	23.910	39.388	46.747	76.705	93.936
	16 × 16	13.214	23.696	39.366	46.747	62.309	64.636
	19 × 19	13.214	23.653	39.366	46.341	60.605	64.636
	20 × 20	13.214	23.645	39.366	46.341	60.497	64.636

From Pradhan, K.K., Chakraverty, S., 2014. Free vibration of functionally graded thin elliptic plates with various edge supports. Structural Engineering and Mechanics 53 (2), 337–354.

circular plate is considered in Table 6.2 and elliptic plate (ratio of semi-major to semi-minor axes; $a/b = 2$) in Table 6.3, respectively, with clamped (C) and simply supported (S) edge supports. Rather than taking natural frequencies for the combination of symmetric and antisymmetric modes separately, the present study computes frequencies for all the modes at a time. This means that earlier authors used the deflection function of odd-odd, even-odd, odd-even, and even-even polynomials of x and y separately. The present authors have taken all the powers of x and y for the ease in computation in a single run. So the number of approximations may be seen less in previous works. It is interesting to note here that an increase in the number of polynomials in displacement component plays a crucial role in the convergence of non-dimensional frequencies, irrespective of geometric configuration and edge support of the plate.

After a satisfactory test of convergence, we have performed a comparison study for non-dimensional frequencies of elliptic (or circular) plates with the existing literature. As there is not much investigation related to free vibration of elliptic FG plates, it is worth taking isotropic plates to compare the results. The first five non-dimensional frequencies of isotropic circular ($a/b = 1.0$) and elliptic plate with different boundary conditions are compared in Table 6.4, assuming Poisson's ratio as $\nu = 0.3$. It may be concluded that frequencies related to the present study are in excellent agreement with the previous results.

Table 6.4 Comparison of the first five non-dimensional frequencies for isotropic elliptic plate

a/b	BCs	Sources	λ_1	λ_2	λ_3	λ_4	λ_5
	C	Present	10.2158	21.261	34.878	39.773	51.209
	C	Exact	10.216	21.260	34.878	39.773	-
	C	Leissa (1969)	10.2158	21.26	34.88	39.771	51.04
	C	Mazumdar (1971)	10.2151	-	-	-	-
	C	Cheung and Tham (1988)	10.2062	21.27	34.94	40.21	52.05
	C	Singh and Chakraverty (1992a, 1994b)	10.216	21.260	34.878	39.773	-
	C	Rajalingham et al. (1994)	10.2158	21.2604	34.8770	39.7711	51.0300
	C	Chakraverty and Petyt (1997)	10.216	21.260	34.878	39.773	51.030
	C	Wu and Liu (2001); Wu et al. (2002)	10.216	21.260	34.877	39.771	51.030
	C	Prakash and Ganpathi (2006)	10.213	21.259	34.849	-	50.974
	C	Chakraverty et al. (2007)	10.2158	21.2604	34.8770	39.7712	-
	S	Present	4.9351	13.899	25.619	29.737	40.915
	S	Exact	4.935	13.898	25.613	29.720	-
	S	Leissa (1969)	4.9351	13.8982	25.6173	29.7200	39.9573
1.0	S	Leissa and Narita (1980)	4.93515	13.8982	25.6133	29.7200	39.9573
	S	Cheung and Tham (1988)	4.927	13.88	25.54	29.84	40.30
	S	Singh and Chakraverty (1992b, 1994b)	4.9351	13.898	25.613	29.720	-
	S	Chakraverty and Petyt (1997)	4.9351	13.898	25.613	29.720	39.957
	S	Wu and Liu (2001); Wu et al. (2002)	4.935	13.898	25.613	29.720	39.957
	S	Prakash and Ganpathi (2006)	4.935	13.898	25.613	-	39.957
	S	Chakraverty et al. (2007)	4.9351	13.8982	25.6133	29.7201	-
	F	Present	5.3583	9.0035	12.5645	21.2331	22.1935
	F	Exact	5.3583	9.0031	12.439	20.475	-

Continued

Table 6.4 Comparison of the first five non-dimensional frequencies for isotropic elliptic plate—cont'd.

a/b	BCs	Sources	λ_1	λ_2	λ_3	λ_4	λ_5
	F	Leissa (1969)	5.253	9.084	12.23	20.52	21.6
	F	Singh and Chakraverty (1991, 1994b)	5.3583	9.0031	12.439	20.475	-
	F	Chakraverty and Petyt (1997)	5.3583	9.0031	12.439	20.475	21.835
	F	Wu and Liu (2001); Wu et al. (2002)	5.358	9.003	12.439	20.475	21.835
	F	Chakraverty et al. (2007)	5.3583	9.0031	12.4390	20.4746	-
	C	Present	27.377	39.499	55.985	69.863	78.020
	C	Leissa (1969)	27.378	-	-	-	-
	C	Mazumdar (1971)	27.741	-	-	-	-
	C	Singh and Chakraverty (1992a)	27.377	39.497	55.985	69.858	-
	C	Singh and Chakraverty (1994b)	27.377	39.497	55.985	69.858	77.037
	C	Chakraverty et al. (2007)	27.3774	39.4974	55.9758	69.8580	-
2.0	S	Present	13.213	23.645	38.354	46.165	60.497
	S	Singh and Chakraverty (1992b)	13.213	23.641	38.354	46.151	57.625
	S	Singh and Chakraverty (1994b)	13.213	23.641	38.354	46.151	-
	S	Chakraverty et al. (2007)	13.2135	23.6410	38.3259	46.1504	-
	F	Present	6.6706	10.548	17.213	22.353	32.696
	F	Singh and Chakraverty (1994b)	6.6706	10.548	16.923	22.019	-
	F	Chakraverty et al. (2007)	6.6705	10.5476	16.9212	22.0149	-

From Pradhan, K.K., Chakraverty, S., 2014. Free vibration of functionally graded thin elliptic plates with various edge supports. Structural Engineering and Mechanics 53 (2), 337–354.

6.3 RESULTS AND DISCUSSIONS

In view of the above verification, the effect of aspect ratios (a/b) on the first six non-dimensional frequencies are reported in Tables 6.5–6.7 with various boundary conditions and gradation of properties in FG constituents. Clamped edge support is considered in Table 6.5 with different power-law exponents (k) to evaluate the first six non-dimensional frequencies of elliptic FG plates. In a similar fashion, frequencies are computed for simply supported and completely free edge supports in Tables 6.6 and 6.7, respectively. In the case of clamped and simply supported edge supports, it can be observed that frequencies are increasing with an increase in aspect ratios and act reverse with an increase in power-law indices, whereas frequencies follow peculiar behavior in the case of completely free FG elliptic plates, that is, frequencies are decreasing with an increase in k and are showing fluctuations with an increase in a/b at higher modes.

In Tables 6.8 and 6.9, the effects of variation of Poisson's ratio (ν) on natural frequencies of elliptic (or circular) FG plates supported by various

Table 6.5 Effect of aspect ratios (a/b) on the lowest frequencies of clamped elliptic FG plates

a/b	k	λ_1	λ_2	λ_3	λ_4	λ_5	λ_6
1.0	0.0	10.216	21.261	34.878	39.773	51.209	61.407
	0.1	9.851	20.501	33.631	38.352	49.379	59.213
	1.0	8.500	17.689	29.020	33.093	42.609	51.094
	2.0	8.125	16.909	27.741	31.634	40.730	48.841
1.5	0.0	17.129	28.472	41.487	44.392	57.038	65.369
	0.1	16.517	27.454	40.005	42.806	54.999	63.032
	1.0	14.253	23.690	34.519	36.936	47.458	54.389
	2.0	13.624	22.646	32.998	35.308	45.366	51.992
2.0	0.0	27.377	39.499	55.985	69.863	78.020	88.074
	0.1	26.399	38.087	53.984	67.366	75.232	84.926
	1.0	22.779	32.865	46.582	58.129	64.917	73.282
	2.0	21.775	31.416	44.529	55.566	62.055	70.051
3.0	0.0	56.801	71.626	90.350	116.81	147.34	150.18
	0.1	54.771	69.066	87.121	112.64	142.08	144.81
	1.0	47.261	59.596	75.176	97.195	122.59	124.95
	2.0	45.177	56.969	71.862	92.909	117.19	119.44

From Pradhan, K.K., Chakraverty, S., 2014. Free vibration of functionally graded thin elliptic plates with various edge supports. Structural Engineering and Mechanics 53 (2), 337–354.

Table 6.6 Effect of aspect ratios (a/b) on the lowest frequencies of simply supported elliptic FG plates

a/b	k	λ_1	λ_2	λ_3	λ_4	λ_5	λ_6
1.0	0.0	4.935	13.899	25.619	29.736	40.915	51.072
	0.1	4.759	13.402	24.703	28.674	39.453	49.247
	1.0	4.106	11.564	21.316	24.742	34.043	42.495
	2.0	3.925	11.055	20.376	23.651	32.542	40.621
1.5	0.0	8.282	17.835	27.452	31.805	41.409	52.246
	0.1	7.986	17.198	26.471	30.668	39.929	50.379
	1.0	6.891	14.839	22.842	26.463	34.454	43.472
	2.0	6.587	14.186	21.835	25.297	32.935	41.555
2.0	0.0	13.213	23.645	38.354	46.165	60.497	62.848
	0.1	12.741	22.799	36.984	44.515	58.335	60.602
	1.0	10.994	19.673	31.913	38.412	50.337	52.292
	2.0	10.509	18.806	30.506	36.718	48.118	49.987
3.0	0.0	27.081	40.146	57.050	84.282	98.673	116.06
	0.1	26.114	38.711	55.011	81.269	95.146	111.91
	1.0	22.533	33.403	47.469	70.127	82.101	96.569
	2.0	21.539	31.931	45.376	67.035	78.481	92.312

From Pradhan, K.K., Chakraverty, S., 2014. Free vibration of functionally graded thin elliptic plates with various edge supports. Structural Engineering and Mechanics 53 (2), 337–354.

Table 6.7 Effect of aspect ratios (a/b) on the lowest frequencies of completely free elliptic FG plates

a/b	k	λ_1	λ_2	λ_3	λ_4	λ_5	λ_6
1.0	0.0	5.358	9.003	12.564	21.233	22.193	37.564
	0.1	5.167	8.682	12.115	20.474	21.400	36.221
	1.0	4.458	7.491	10.454	17.667	18.466	31.255
	2.0	4.262	7.161	9.993	16.888	17.652	29.877
1.5	0.0	6.477	7.986	16.309	16.510	17.767	29.971
	0.1	6.245	7.701	15.726	15.920	17.132	28.899
	1.0	5.389	6.645	13.569	13.737	14.783	24.937
	2.0	5.151	6.352	12.971	13.132	14.131	23.838
2.0	0.0	6.671	10.548	17.212	22.353	27.773	32.696
	0.1	6.432	10.171	16.596	21.554	26.780	31.527
	1.0	5.550	8.776	14.321	18.599	23.108	27.205
	2.0	5.306	8.389	13.689	17.779	22.089	26.005
3.0	0.0	6.757	15.615	17.618	31.415	33.961	51.246
	0.1	6.516	15.057	16.988	30.292	32.748	49.414
	1.0	5.622	12.992	14.659	26.139	28.258	42.639
	2.0	5.374	12.419	14.013	24.986	27.012	40.759

From Pradhan, K.K., Chakraverty, S., 2014. Free vibration of functionally graded thin elliptic plates with various edge supports. Structural Engineering and Mechanics 53 (2), 337–354.

Table 6.8 Effect of Poisson's ratio (ν) on non-dimensional frequencies of elliptic FG plates with various BCs for $k = 0$

a/b	BC	ν	λ_1	λ_2	λ_3	λ_4	λ_5	λ_6
1.0	C	0.00	10.216	21.260	34.878	39.773	51.209	61.407
		0.25	10.216	21.260	34.878	39.773	51.209	61.407
		0.33	10.216	21.260	34.878	39.773	51.209	61.407
		0.50	10.216	21.260	34.878	39.773	51.209	61.407
	S	0.00	4.4436	13.502	25.249	29.379	40.519	50.644
		0.25	4.8601	13.835	25.559	29.678	40.851	51.002
		0.33	4.9790	13.936	25.654	29.771	40.953	51.114
		0.50	5.2127	14.141	25.849	29.963	41.166	51.347
	F	0.00	6.1531	8.2441	14.148	20.758	24.659	37.779
		0.25	5.5112	8.8902	12.881	21.158	22.701	37.599
		0.33	5.2620	9.0692	12.363	21.277	21.867	37.543
		0.50	4.6404	9.4141	11.021	19.648	21.519	34.641
2.0	C	0.00	27.377	39.499	55.985	69.862	78.019	88.074
		0.25	27.377	39.499	55.985	69.862	78.019	88.074
		0.33	27.377	39.499	55.985	69.862	78.019	88.074
		0.50	27.377	39.499	55.985	69.862	78.019	88.074
	S	0.00	12.646	22.829	37.367	45.803	59.139	62.416
		0.25	13.125	23.519	38.201	46.106	60.283	62.777
		0.33	13.265	23.718	38.444	46.201	60.624	62.889
		0.50	13.546	24.115	38.930	46.399	61.314	63.126
	F	0.00	7.060	12.269	18.235	25.329	27.675	34.809
		0.25	6.778	10.870	17.473	22.938	27.775	33.196
		0.33	6.597	10.346	17.034	21.981	27.768	32.359
		0.50	6.037	9.0676	15.683	19.540	27.710	29.828

From Pradhan, K.K., Chakraverty, S., 2014. Free vibration of functionally graded thin elliptic plates with various edge supports. Structural Engineering and Mechanics 53 (2), 337–354.

edge conditions with fixed aspect ratio (a/b) and power-law index (k) are summarized. For both isotropic ($k = 0$) and FG plates, it can be easily seen that frequencies remain constant for different values of ν in the case of clamped plates. Assuming simply supported, frequencies are increasing with an increase in ν, whereas frequencies are showing fluctuating order while considering free edge condition, keeping both a/b and k fixed. One may conclude that effects of edge supports and Poisson's ratio on free vibration response of FG elliptic plates is also quite similar to that of isotropic elliptic plates.

Table 6.9 Effect of Poisson's ratio (ν) on non-dimensional frequencies of elliptic FG plates with various BCs for $k = 1$

a/b	BC	ν	λ_1	λ_2	λ_3	λ_4	λ_5	λ_6
1.0	C	0.00	8.5001	17.689	29.019	33.093	42.609	51.094
		0.25	8.5001	17.689	29.019	33.093	42.609	51.094
		0.33	8.5001	17.689	29.019	33.093	42.609	51.094
		0.50	8.5001	17.689	29.019	33.093	42.609	51.094
	S	0.00	3.6973	11.234	21.009	24.445	33.715	42.139
		0.25	4.0439	11.512	21.266	24.694	33.989	42.436
		0.33	4.1428	11.596	21.346	24.771	34.075	42.529
		0.50	4.3372	11.766	21.508	24.931	34.253	42.724
	F	0.00	5.1197	6.8595	11.772	17.272	20.517	31.434
		0.25	4.5856	7.3971	10.718	17.605	18.889	31.284
		0.33	4.3783	7.5461	10.286	17.704	18.194	31.238
		0.50	3.8610	7.8330	9.1701	16.348	17.9045	28.823
2.0	C	0.00	22.779	32.865	46.582	58.129	64.916	73.282
		0.25	22.779	32.865	46.582	58.129	64.916	73.282
		0.33	22.779	32.865	46.582	58.129	64.916	73.282
		0.50	22.779	32.865	46.582	58.129	64.916	73.282
	S	0.00	10.522	18.995	31.092	38.111	49.207	51.933
		0.25	10.921	19.569	31.785	38.362	50.158	52.234
		0.33	11.038	19.735	31.988	38.441	50.442	52.328
		0.50	11.271	20.065	32.392	38.607	51.017	52.524
	F	0.00	5.874	10.208	15.173	21.075	23.027	28.962
		0.25	5.639	9.0445	14.538	19.085	23.111	27.621
		0.33	5.489	8.6086	14.173	18.289	23.105	26.925
		0.50	5.023	7.5447	13.049	16.258	23.056	24.819

From Pradhan, K.K., Chakraverty, S., 2014. Free vibration of functionally graded thin elliptic plates with various edge supports. Structural Engineering and Mechanics 53 (2), 337–354.

Three-dimensional mode shapes of FG circular and elliptic plates are plotted in Figs. 6.1–6.3, taking various BCs with $k = 1$. Figure 6.1 is meant for clamped FG plates. In a similar fashion, Fig. 6.2 is for simply supported and Fig. 6.3 is for completely free FG plates, respectively. Looking into the deflected shapes of mode shapes, it is easy to predict the edge support to be clamped, simply supported, or free, irrespective of the geometry and power-law exponent used in gradation.

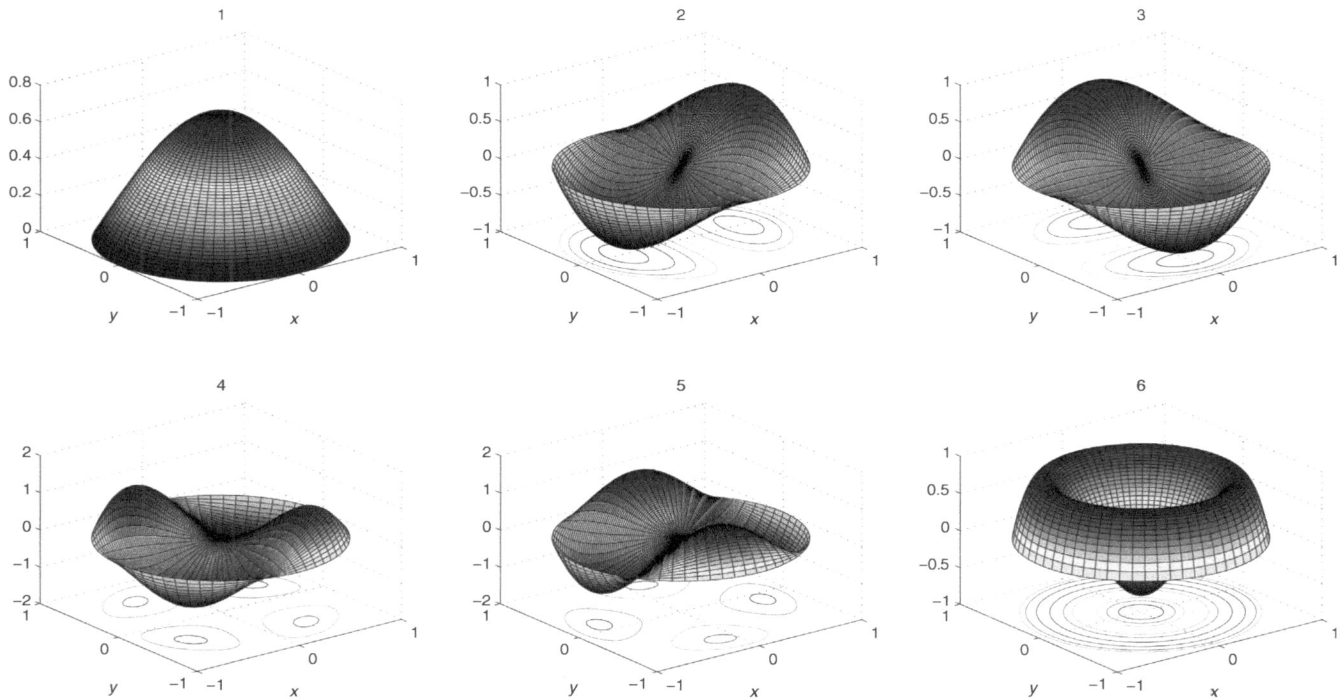

(a) FG circular plate

Figure 6.1 Three-dimensional mode shapes of clamped (a) circular and

(Continued)

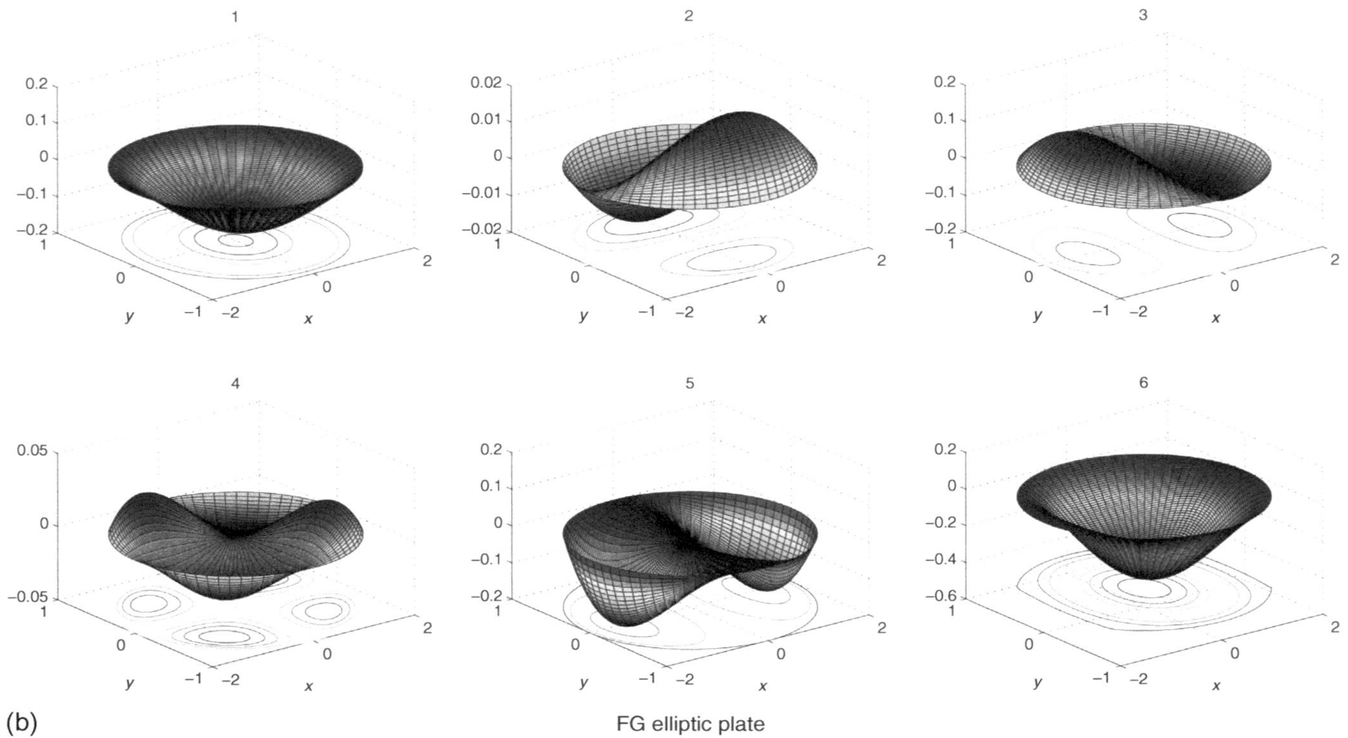

Figure 6.1, cont'd (b) elliptic FG plates with $k = 1$. (*From Pradhan, K.K., Chakraverty, S., 2014. Free vibration of functionally graded thin elliptic plates with various edge supports. Structural Engineering and Mechanics 53 (2), 337–354*).

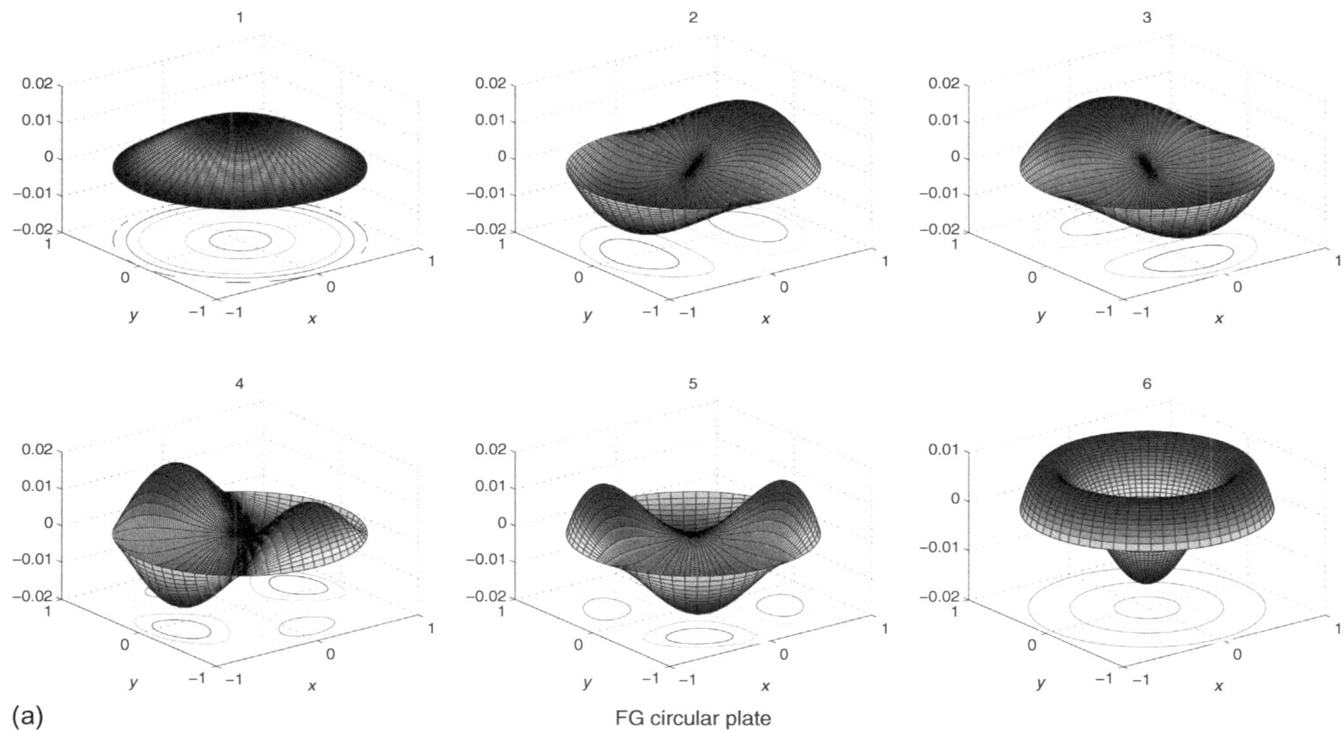

Figure 6.2 3D mode shapes of simply supported (a) circular and

(Continued)

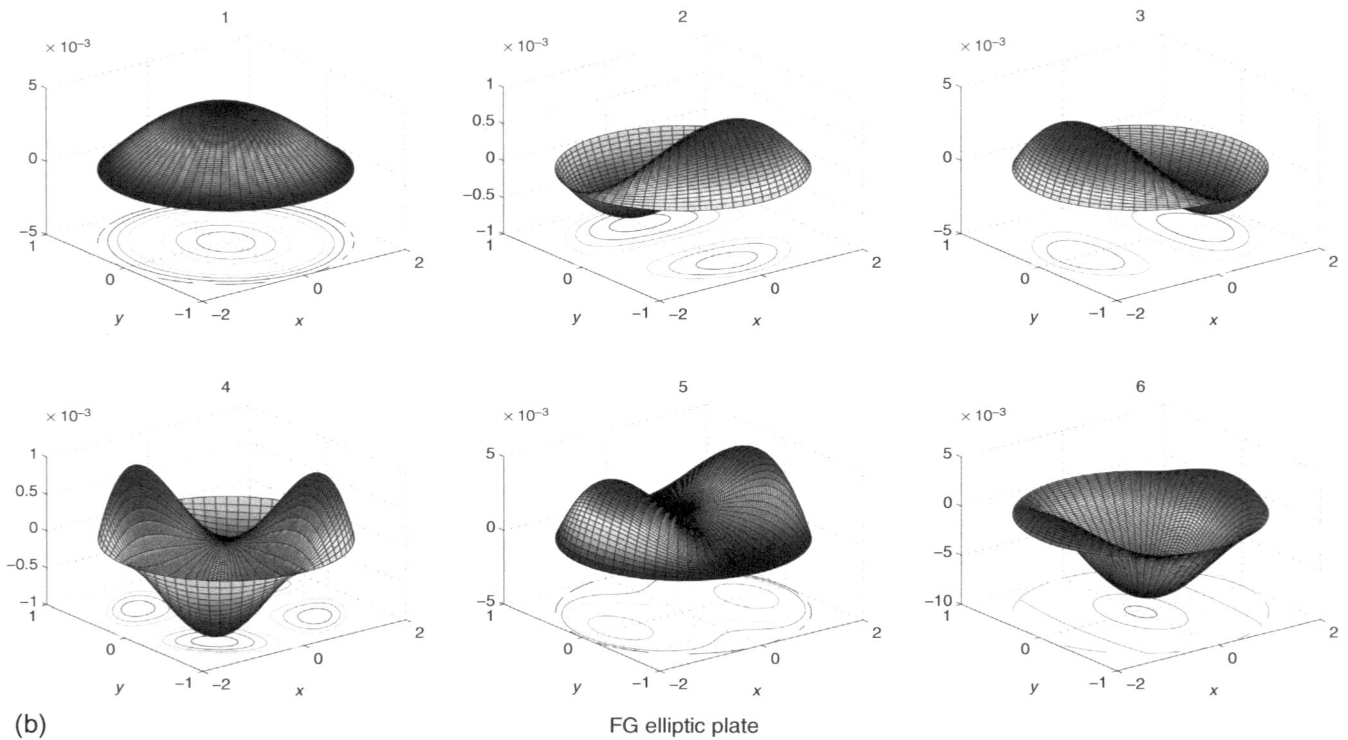

Figure 6.2, cont'd (b) elliptic FG plates with $k = 1$.(*From Pradhan, K.K., Chakraverty, S., 2014. Free vibration of functionally graded thin elliptic plates with various edge supports. Structural Engineering and Mechanics 53 (2), 337–354*).

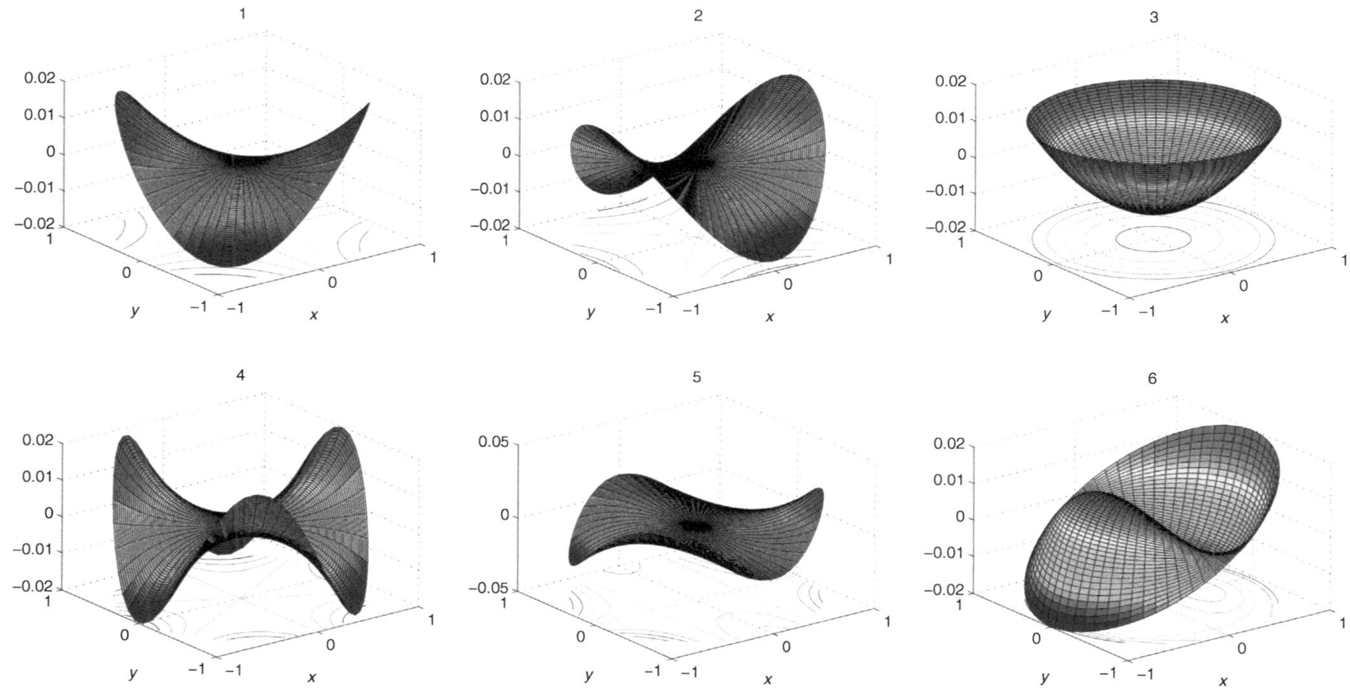

(a)

FG circular plate

Figure 6.3 3D mode shapes of free (a) circular and

(Continued)

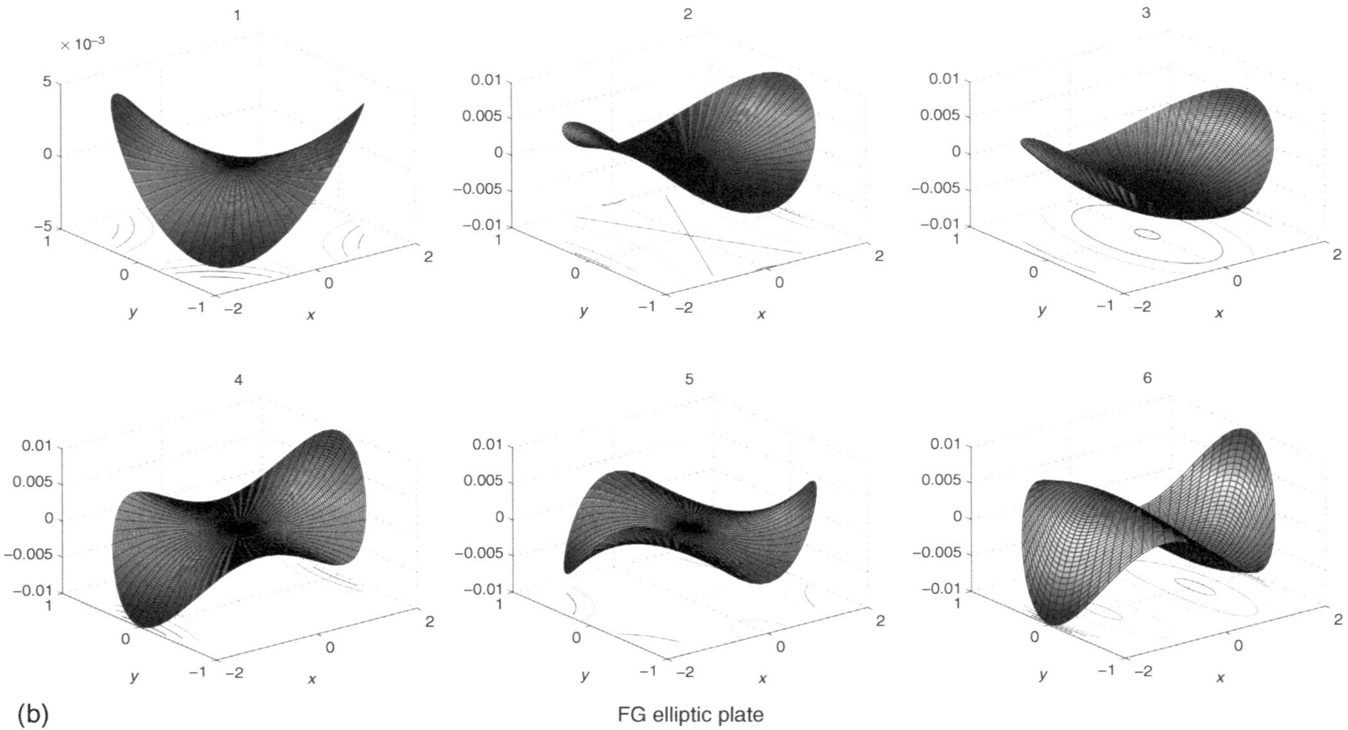

Figure 6.3, cont'd (b) elliptic FG plates with $k = 1$. (*From Pradhan, K.K., Chakraverty, S., 2014. Free vibration of functionally graded thin elliptic plates with various edge supports. Structural Engineering and Mechanics 53 (2), 337–354*).

6.4 CONCLUDING REMARKS

The natural frequencies and mode shapes of FG elliptic and circular plates with various boundary conditions based on classical plate theory were evaluated in this investigation. The generalized eigenfrequency equation for free vibration can be obtained by means of the Rayleigh-Ritz method. Trial functions denoting displacement component may handle any classical boundary condition very easily. Looking into the numerical formulation and results, one may conclude as follows.

- The aspect ratios (a/b), power-law indices (k), and different material distributions are key factors when studying free vibration characteristics of FG elliptic (or circular) plates.
- In the Rayleigh-Ritz method, an increase in the number of polynomials (n) plays a crucial role in the convergence of non-dimensional frequencies.
- From the tables and figures in this chapter, it may be observed that the frequencies are increasing with an increase in aspect ratios for a fixed power-law index and are decreasing with an increase in power-law exponents for a fixed aspect ratio. This behavior is true for clamped and simply supported edge supports, whereas ambiguity occurs with the effect of aspect ratio on frequencies in completely free FG plates.
- Assuming the effect of Poisson's ratio (ν), one may see that frequencies are independent of ν for clamped elliptic plates. But frequencies are increasing with an increase in ν in the case of simply supported and are showing fluctuating behavior while considering the free edge condition, keeping both a/b and k fixed.
- Other shear deformation plate theories can also be extended easily following the above analysis.

CHAPTER 7

Vibration Problems of Functionally Graded Triangular Plates

In structural design and architecture, it may also be observed that triangular plates have very important applications rather than only rectangular and elliptic shapes. On this basis, literature survey in this part provides the research efforts on isotropic triangular plates and a very few investigations performed in functionally graded (FG) triangular plates. In this regard, Mirza and Bijlani (1985) have found the natural frequencies and mode shapes of cantilevered triangular plates with variable thickness using the finite element technique. Gorman (1983, 1986, 1989) has proposed a highly accurate analytical solution (the method of superposition) for free vibration of right triangular plates with simply supported edge supports, with combinations of clamped-simply supported boundary supports and different boundary conditions (BCs) with one edge free, respectively. A highly accurate simplified solution has been given by Saliba (1990) to study the free vibration of simply supported right triangular thin plates. Singh and Chakraverty (1992c) have done an exhaustive study on isotropic triangular plates by applying the Rayleigh-Ritz method to handle all sets of classical edge supports. Wanji and Cheung (1998) have found the bending, vibration, and buckling of a refined triangular discrete Kirchoff thin plate element (RDKT) in order to improve the results for the original triangular discrete Kirchoff thin plate element (DKT). An approximate method using the Green function is developed by Sakiyama and Huang (2000) for free vibration of right triangular plates with variable thickness. Zhong (2000) has applied a triangular differential quadrature method to study free flexural vibration of isosceles triangular Mindlin plates. Reddy's third-order plate theory is used by Cheng and Batra (2000) to study buckling and steady state vibrations of a simply supported FG polygonal plate resting on a Winkler-Pasternak elastic foundation. Non-dimensional influence (Green's) function has been applied by Kang and Lee (2001) in free vibration analysis of

http://dx.doi.org/10.1016/B978-0-12-804228-1.00007-9

arbitrarily shaped plates with clamped edges. Free vibration of cantilevered and completely free isosceles triangular plates has been investigated by Cheung and Zhou (2002) based on exact three-dimensional elasticity theory. Moreover, very few authors have worked especially on FG triangular plates. Belalia and Houmat (2012) have developed a p-version finite element method based on a curved triangular p-element to analyze nonlinear free vibration of FG sector plates.

On this note, this chapter deals with free vibration of FG thin triangular plates subject to various classical BCs at the three edges. Here, four different types of triangular plates are considered, viz. two right-angled, an equilateral, and an isosceles. The power-law variation of FG material properties is also considered. The plate displacement components are expressed as the linear combination of simple algebraic polynomials (generated from Pascal's triangle) and the Rayleigh-Ritz procedure is applied to generate the generalized eigenvalue problem. The triangular plate generally considers coordinate transformation to standard triangle and leads to a slight modification in the energy expressions in the Rayleigh-Ritz method. Corresponding natural frequencies are incorporated after test of convergence and verification with existing results in special cases. Three-dimensional mode shapes for clamped-clamped-clamped (C-C-C) and clamped-free-free (C-F-F) triangular plates are also depicted corresponding to the first six non-dimensional frequencies.

7.1 TYPES OF FG TRIANGULAR ELEMENTS

In the cartesian coordinate system, a given FG triangle can be completely determined by three numbers, a, b, and c, as shown in Fig. 7.1. Let us now apply a transformation of coordinates from Cartesian (x, y) to natural coordinate system (ξ, η) to generate a standard triangle as given below (Singh and Chakraverty, 1992c):

$$\xi = \frac{\left(x - \frac{by}{c}\right)}{a}, \quad \eta = \frac{y}{c}; \quad x = a\xi + b\eta, \quad y = c\eta \tag{7.1}$$

One may notice, as mentioned in Singh and Chakraverty (1992c), that unlike finite element methods where one element is transformed into another element of similar shape, we use Eq. (7.1) globally for the whole plate.

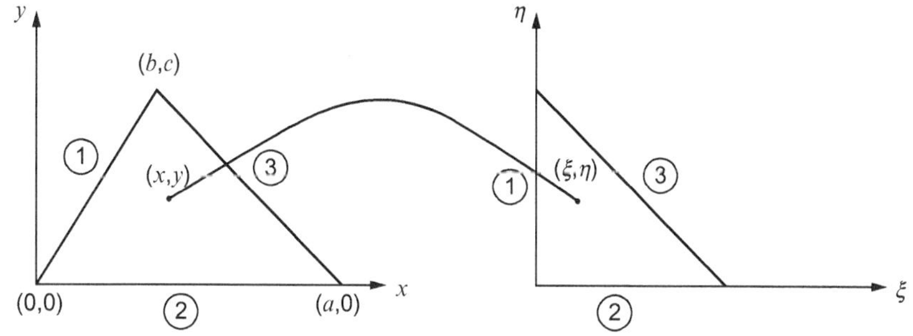

Figure 7.1 Coordinate transformation of given generalized triangle onto the standard triangle.

For the present investigation, we have considered four special cases in the same fashion in Singh and Chakraverty (1992c) for FG triangular plates as represented in Fig. 7.2. Four cases as depicted in Fig. 7.2(a)–(d) have the following features:

- The right-angled triangle is considered in Fig. 7.2(a) with $\theta = 0$ and different $\mu = 1.0, 1.5, 2.0, 2.5, 3.0$. The lengths of the sides 1, 2, and 3 are μ, 1 and $\sqrt{(1+\mu^2)}$, respectively.
- Figure 7.2(b) shows the equilateral triangle with each sides as unity in non-dimensional form. Other geometrical parameters for this triangle are $\theta = 1/\sqrt{3}$ and $\mu = \sqrt{3}/2$.
- A right-angled triangle with angles 30°, 60°, and 90° is demonstrated in Fig. 7.2(c). Sides 1, 2, and 3 have the lengths as $1/\sqrt{3}$, 1, and $2/\sqrt{3}$, respectively. Accordingly, we can find the values of $\theta = 0$ and $\mu = 1/\sqrt{3}$.
- Lastly, Fig. 7.2(d) represents an isosceles triangle with angles 30°, 30°, and 120°. The non-dimensionalized lengths of the sides 1, 2, and 3 are 1, 1, and $\sqrt{3}$, respectively, with $\theta = -1/\sqrt{3}$ and $\mu = \sqrt{3}/2$.

In these figures, the BCs may very well be controlled by the assigned edges 1, 2, and 3, respectively.

7.2 NUMERICAL MODELING

In the case of FG triangular plates, there occurs transformation of geometry from general cartesian coordinate system to standard triangle. In this regard, the numerical modeling needs a few modifications in finding the generalized eigenvalue problem associated with free vibration problem.

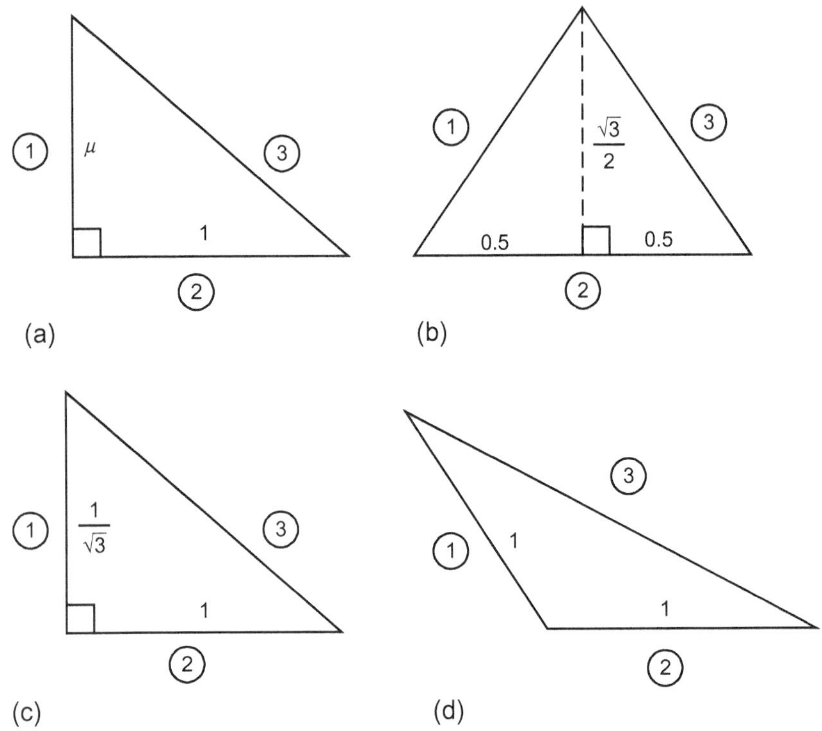

Figure 7.2 Four different cases of functionally graded triangular plates. (a) Right-angled triangle with different μ. (b) Equilateral triangle. (c) Right-angled triangle with angles 30°, 60°, and 90°. (d) Isosceles triangle with angles 30°, 30°, and 120°.

Accordingly, we may find the strain energy $\mathbf{U}$ and kinetic energy $\mathbf{T}$ of the plate in the form of Eqs. (3.19) and (3.20), respectively. Assuming the coordinate transformation given in Fig. 7.1, the expression for maximum strain and kinetic energies can be written as

$$\mathbf{U}_{\max} = \frac{1}{2}\int_{\Omega}\left[D_{11}\left\{\left(\frac{\partial^2 W}{\partial \xi^2}\right)^2 + \left(\frac{\partial^2 W}{\partial \eta^2}\right)^2\right\} + 2D_{12}\frac{\partial^2 W}{\partial \xi^2}\frac{\partial^2 W}{\partial \eta^2} + 4D_{66}\left(\frac{\partial^2 W}{\partial \xi \partial \eta}\right)^2\right]|J(\xi,\eta)|\, d\xi\, d\eta \tag{7.2}$$

$$\mathbf{T}_{\max} = \frac{\omega^2}{2}\int_{\Omega} I_0 W^2\, |J(\xi,\eta)|\, d\xi\, d\eta \tag{7.3}$$

In Eqs. (7.2) and (7.3), $|J(\xi,\eta)|$ represents the Jacobian resulted for the coordinate transformation from (x, y) to (ξ, η) and is given by

$$J(\xi,\eta) = \frac{\partial(x,y)}{\partial(\xi,\eta)} = \begin{pmatrix} \frac{\partial x}{\partial \xi} & \frac{\partial x}{\partial \eta} \\ \frac{\partial y}{\partial \xi} & \frac{\partial y}{\partial \eta} \end{pmatrix} = \begin{pmatrix} a & b \\ 0 & c \end{pmatrix} = ac$$

Application of the Rayleigh-Ritz Method

The amplitude $W(\xi,\eta)$ may be expressed as the sum of series of simple algebraic polynomials involving both ξ and η as follows:

$$W(\xi,\eta) = \sum_{i=1}^{n} c_i \varphi_i(\xi,\eta)$$

where c_i are unknown constants to be determined and φ_i are the admissible functions, which satisfy the essential BCs and are represented as

$$\varphi_i(\xi,\eta) = f\psi_i(\xi,\eta), \quad i = 0, 1, 2, \ldots, n$$

Here, n is the number of polynomials involved in the admissible functions. The function $f = \xi^p \eta^q (1-\xi-\eta)^r$ with the exponents p, q, and r, which control various BCs. The parameter $p = 0$, 1, or 2 according as the side $\xi = 0$ is free (F), simply supported (S) or clamped (C). Similar interpretations can be given to the parameters q and r corresponding to the sides $\eta = 0$ and $\xi + \eta = 1$, respectively. The components of ψ_i may be generated from Pascal's triangle as given in Table 7.1.

Assuming constant Poisson's ratio (ν), the Rayleigh quotient is obtained by equating $\mathbf{U}_{\max}$ and $\mathbf{T}_{\max}$ as

$$\omega^2 = \frac{\int_\Omega D_{11}\left[\left\{\left(\frac{\partial^2 W}{\partial \xi^2}\right)^2 + \left(\frac{\partial^2 W}{\partial \eta^2}\right)^2\right\} + 2\nu \frac{\partial^2 W}{\partial \xi^2}\frac{\partial^2 W}{\partial \eta^2} + 2(1-\nu)\left(\frac{\partial^2 W}{\partial \xi \partial \eta}\right)^2\right] d\xi\, d\eta}{\int_\Omega I_0 W^2\, d\xi\, d\eta} \tag{7.4}$$

Table 7.1 Ten algebraic polynomials obtained from Pascal's triangle

i	1	2	3	4	5	6	7	8	9	10
ψ_i	1	ξ	η	ξ^2	$\xi\eta$	η^2	ξ^3	$\xi^2\eta$	$\xi\eta^2$	η^3

Next we take the partial derivatives(s) of ω^2 with respect to unknown constants as follows

$$\frac{\partial \omega^2}{\partial c_i} = 0, \quad i = 1, 2, 3, \ldots, n \tag{7.5}$$

Further manipulation of Eq. (7.5) yields the generalized eigenvalue problem of the form

$$\sum_{j=1}^{n} (a_{ij} - \lambda^2 b_{ij}) c_j = 0, \quad i = 1, 2, 3, \ldots, n \tag{7.6}$$

where

$$a_{ij} = \left[12\left(1 - \frac{1}{E_r}\right) \left\{ \frac{1}{k+3} - \frac{1}{k+2} + \frac{1}{4(k+1)} \right\} + \frac{1}{E_r} \right] \int_{\Omega} \left[A_1 \varphi_i^{\xi\xi} \varphi_j^{\xi\xi} + A_2 \varphi_i^{\eta\eta} \varphi_j^{\eta\eta} + A_3 \varphi_i^{\xi\eta} \varphi_j^{\xi\eta} + A_4 \left(\varphi_i^{\xi\xi} \varphi_j^{\xi\eta} + \varphi_i^{\xi\eta} \varphi_j^{\xi\xi} \right) + A_5 \left(\varphi_i^{\xi\eta} \varphi_j^{\eta\eta} + \varphi_i^{\eta\eta} \varphi_j^{\xi\eta} \right) + A_6 \left(\varphi_i^{\xi\xi} \varphi_j^{\eta\eta} + \varphi_i^{\eta\eta} \varphi_j^{\xi\xi} \right) \right] \mathrm{d}\xi \, \mathrm{d}\eta$$

$$b_{ij} = \left\{ \frac{(1 - 1/\rho_r)}{k+1} + \frac{1}{\rho_r} \right\} \int_{\Omega} \varphi_i \varphi_j \, \mathrm{d}\xi \, \mathrm{d}\eta, \quad \lambda^2 = \omega^2 a^4 \rho_c h / D_c$$

with $E_r = E_c/E_m$ and $\rho_r = \rho_c/\rho_m$ being the ratios of Young's moduli and mass densities of FG constituents, respectively, and $D_c = E_c h^3/12(1 - \nu^2)$ being meant for the flexural rigidity of the ceramic constituent of the plate. In the expression of a_{ij}, the superscripted notations stand for partial differentiation and $A_1, A_2, \ldots, A_6$ are defined as

$$A_1 = \left(1 + \theta^2\right)^2, \quad A_2 = \frac{1}{\mu^4}, \quad A_3 = \frac{2\left(1 - \nu + 2\theta^2\right)}{\mu^2}$$

$$A_4 = \frac{-2\theta\left(1 + \theta^2\right)}{\mu}, \quad A_5 = \frac{-2\theta}{\mu^3}, \quad A_6 = \frac{\nu + \theta^2}{\mu^2}$$

with $\theta = \frac{b}{c}, \quad \mu = \frac{c}{a}$

In the eigenvalue problem (Eq. (7.6)), λ and $c_j = [c_1, c_2, c_3, \ldots, c_n]^T$ refer to the non-dimensional frequency (or frequency parameter) and the column vector of unknown constants, respectively. Consequently, we have computed free vibration characteristics for different types of FG triangular plates along with 3D mode shapes in subsequent discussions.

7.3 CONVERGENCE AND COMPARISON STUDIES

The physical parameters that mainly influence the free vibration of FG plates are power-law indices (k), E_r, and ρ_r. Along with these parameters, two geometric parameters, viz. θ and μ, also play crucial roles in these computations. As a result, this section involves the convergence of non-dimensional frequencies for one of the FG triangular plates (Fig. 7.2(a)) and the verification with the available results in special cases.

The convergences of the six lowest non-dimensional frequencies of C-C-C FG triangular plates (Fig. 7.2(a)) are shown in Table 7.2 for different values of power-law index (k), E_r, and ρ_r. The respective values of θ and μ are assumed as 0 and 1. It is evident from this table that convergence of non-dimensional frequencies occurs with the increase in the number of polynomials involved in displacement component, irrespective of the other parameters and edge condition. It can also be noticed here that FG triangular plates yield same results for $k = 0$ and for combined effects of $E_r = 1.0$ and $\rho_r = 1.0$. Hence, isotropic plate results can be obtained from FG plates either by assuming k as 0 or for simultaneous effects of E_r and ρ_r as unity.

In view of the convergences of non-dimensional frequencies, Tables 7.3–7.5 are meant for the comparison of present results with the available results in special cases ($k = 0$; FG plate behaves as isotropic). As no literature is available based on the vibration analysis of FG triangular plates, we have carried out the validation with the results of isotropic plates. The fundamental non-dimensional frequencies of isotropic right-angled triangular plate is considered in Table 7.3 with different values of $\mu = 1.0$, 1.5, 2.0, 2.5, and 3.0. In a similar fashion, the first five non-dimensional frequencies of the right-angled and equilateral triangular plates mentioned respectively in Fig. 7.2(b) and (c) are assumed in Table 7.4. On the other hand, Table 7.5 considers the isosceles triangular plate with angles 30°, 30°, and 120° for the comparison of the first six non-dimensional frequencies. In these tabulations, it can easily be seen that present results are in excellent agreement with the existing literature.

7.4 RESULTS AND DISCUSSIONS

In view of the above test of convergence and comparison studies, we have now evaluated the first six non-dimensional frequencies of FG triangular plates included in Fig. 7.2(a)–(d) for different combination of BCs. As such,

Table 7.2 Convergences of the first six non-dimensional frequencies of C-C-C FG triangular plates with various parameters; $\theta = 0$ and $\mu = 1$

Parameter(s)	Value	n	λ_1	λ_2	λ_3	λ_4	λ_5	λ_6
k ($E_r = 2.0$, $\rho_r = 1.0$)	0	5	94.4970	179.7568	222.8206	305.4952	421.2256	-
		8	94.2693	164.0299	209.9348	291.1884	334.5662	442.9953
		10	94.1526	159.8475	202.0756	290.8454	328.9007	371.6805
		13	93.8589	159.6815	196.7425	255.4264	310.7635	363.8529
		15	93.8094	159.5177	196.4910	248.4158	299.3326	354.9414
	0.1	5	91.7145	174.4639	216.2597	296.4999	408.8226	-
		8	91.4936	159.2000	203.7533	282.6144	324.7150	429.9514
		10	91.3803	155.1408	196.1256	282.2815	319.2162	360.7364
		13	91.0953	154.9797	190.9494	247.9054	301.6131	353.1393
		15	91.0472	154.8207	190.7053	241.1012	290.5188	344.4903
	2.0	5	79.0618	150.3953	186.4251	255.5956	352.4226	-
		8	78.8714	137.2372	175.6441	243.6257	279.9182	370.6365
		10	78.7737	133.7380	169.0686	243.3387	275.1780	310.9702
		13	78.5280	133.5991	164.6066	213.7051	260.0034	304.4212
		15	78.4866	133.4621	164.3961	207.8396	250.4396	296.9653
E_r ($k = 1.0$, $\rho_r = 1.0$)	1.0	5	94.4970	179.7568	222.8206	305.4952	421.2256	-
		8	94.2693	164.0299	209.9348	291.1884	334.5662	442.9953
		10	94.1526	159.8475	202.0756	290.8454	328.9007	371.6805
		13	93.8589	159.6815	196.7425	255.4264	310.7635	363.8529
		15	93.8094	159.5177	196.4910	248.4158	299.3326	354.9414

	2.0	5	81.8368	155.6740	192.9683	264.5666	364.7920	–
		8	81.6396	142.0540	181.8089	252.1766	289.7428	383.6452
		10	81.5385	138.4320	175.0026	251.8795	284.8363	321.8848
		13	81.2842	138.2882	170.3840	221.2057	269.1291	315.1059
		15	81.2413	138.1464	170.1662	215.1344	259.2296	307.3883
	2.5	5	79.0618	150.3953	186.4251	255.5956	352.4226	–
		8	78.8714	137.2372	175.6441	243.6257	279.9182	370.6365
		10	78.7737	133.7380	169.0686	243.3387	275.1780	310.9702
		13	78.5280	133.5991	164.6066	213.7051	260.0034	304.4212
		15	78.4866	133.4621	164.3961	207.8396	250.4396	296.9653
ρ_r ($k = 1.0$, $E_r = 1.0$)	1.0	5	94.4970	179.7568	222.8206	305.4952	421.2256	–
		8	94.2693	164.0299	209.9348	291.1884	334.5662	442.9953
		10	94.1526	159.8475	202.0756	290.8454	328.9007	371.6805
		13	93.8589	159.6815	196.7425	255.4264	310.7635	363.8529
		15	93.8094	159.5177	196.4910	248.4158	299.3326	354.9414
	2.0	5	109.1157	207.5653	257.2911	352.7554	486.3894	–
		8	108.8528	189.4054	242.4119	336.2354	386.3238	511.5269
		10	108.7180	184.5760	233.3369	335.8393	379.7818	429.1797
		13	108.3790	184.3843	227.1786	294.9410	358.8388	420.1412
		15	108.3217	184.1952	226.8882	286.8458	345.6395	409.8511
	2.5	5	112.9455	214.8505	266.3215	365.1366	503.4608	–
		8	112.6734	196.0532	250.9201	348.0367	399.8831	529.4807
		10	112.5338	191.0543	241.5266	347.6267	393.1115	444.2432
		13	112.1829	190.8559	235.1522	305.2929	371.4334	434.8874
		15	112.1237	190.6601	234.8516	296.9136	357.7708	424.2362

Table 7.3 Comparison of the fundamental non-dimensional frequencies of isotropic right-angled triangular plates with different μ

Sides 1-2-3	Source	$\mu = c/a$ 1.0	1.5	2.0	2.5	3.0
C–C–C	Present	93.8002	65.4639	53.4671	46.9018	42.7799
	Kim and Dickinson (1990)	93.79	65.46	53.45	46.89	42.76
	Singh and Chakraverty (1992c)	93.80	65.46	53.46	46.91	42.81
	Singh and Saxena (1996)	93.791	-	53.448	-	-
F–S–C	Present	31.7854	19.3552	14.4568	11.9106	10.3690
	Kim and Dickinson (1990)	31.78	19.35	14.46	11.91	10.36
	Singh and Chakraverty (1992c)	31.78	19.35	14.46	11.93	10.39
	Singh and Saxena (1996)	31.784	-	14.455	-	-
F–C–F	Present	6.1730	2.8786	1.6581	1.0753	0.7528
	Kim and Dickinson (1990)	6.168	2.876	1.656	1.074	0.7514
	Singh and Chakraverty (1992c)	6.173	2.878	1.658	1.075	0.7528
	Singh and Saxena (1996)	6.1693	-	1.6569	-	-

Table 7.4 Comparison of the first five non-dimensional frequencies of isotropic triangular plates mentioned in Fig. 7.2(b) and (c)

Sides 1-2-3	Plate mentioned in	Source	λ_1	λ_2	λ_3	λ_4	λ_5
C-C-C	Fig. 7.2(b)	Present	99.0221	189.0510	189.0510	296.8529	316.8320
		Singh and Chakraverty (1992c)	99.022	189.05	189.22	296.85	316.83
		Singh and Saxena (1996)	99.020	189.02	189.22	-	-
	Fig. 7.2(c)	Present	176.5822	280.9494	380.8706	416.4873	558.4695
		Singh and Chakraverty (1992c)	176.58	280.95	380.92	416.53	558.48
		Singh and Saxena (1996)	176.58	279.80	380.06	-	-
C-C-S	Fig. 7.2(b)	Present	81.6027	165.1243	165.3528	268.1460	286.8021
		Singh and Chakraverty (1992c)	81.604	165.12	155.52	271.30	286.95
		Singh and Saxena (1996)	81.601	165.00	155.33	-	-
	Fig. 7.2(c)	Present	137.0544	231.2049	323.9988	362.3161	493.2550
		Singh and Chakraverty (1992c)	137.05	231.20	324.00	362.31	493.97
		Singh and Saxena (1996)	137.02	230.52	321.25	-	-
F-F-S	Fig. 7.2(b)	Present	22.6596	26.6623	70.6584	74.2863	91.5426
		Singh and Chakraverty (1992c)	22.666	26.717	71.033	74.867	91.959
		Singh and Saxena (1996)	22.646	26.661	69.488	-	-
	Fig. 7.2(c)	Present	31.8430	68.7922	88.4578	150.3555	167.7079
		Singh and Chakraverty (1992c)	31.843	68.904	88.635	150.70	167.90
		Singh and Saxena (1996)	31.836	68.252	86.485	-	-
F-C-F	Fig. 7.2(b)	Present	8.9216	35.1313	38.5031	90.5857	93.5195
		Bhat (1987)	8.9221	35.132	38.505	90.590	93.525
		Singh and Chakraverty (1992c)	8.9219	35.155	38.503	91.624	96.725
		Singh and Saxena (1996)	8.9208	35.105	38.487	-	-
	Fig. 7.2(c)	Present	16.9606	49.7370	87.6047	108.4652	180.4222
		Singh and Chakraverty (1992c)	16.960	49.737	87.604	108.46	180.78
		Singh and Saxena (1996)	16.948	49.712	87.338	-	-

Table 7.5 Comparison of first six non-dimensional frequencies of isotropic isosceles triangular plates with angles 30°, 30°, and 120°

Sides 1-2-3	Source	λ_1	λ_2	λ_3	λ_4	λ_5	λ_6
F-C-F	Present	5.7167	21.5246	37.4540	56.0597	74.6846	120.6420
	Mirza and Bijlani (1985)	5.7667	21.100	35.950	54.143	67.425	102.56
	Bhat (1987)	5.7170	21.525	37.455	56.061	74.625	120.65
	Singh and Chakraverty (1992c)	5.7167	21.524	37.456	56.141	74.769	121.16

we have incorporated the effect of three major physical parameters, viz. k, E_r, and ρ_r in three subsections, as discussed in Section 7.3.

7.4.1 Effect of Power-Law Indices (k)

The variation of the first six non-dimensional frequencies of the right-angled FG triangular element (Fig. 7.2(a)) with an increase in k are computed in Tables 7.6–7.11 and subjected to six different boundary supports, viz. C-C-C, C-C-S, S-S-S, C-C-F, C-S-F, and C-F-F, and also for different values of $\mu = 1.0$, 1.5, 2.0, 2.5, and 3.0. Similar results for the FG triangular plates of Fig. 7.2(b)–(d) are reported in Tables 7.12–7.14, respectively. Ten different combination of edge supports, viz. C-C-C, C-C-S, C-C-F, S-S-S, S-C-S, S-C-F, S-F-S, F-F-F, F-C-F, and F-F-S, are considered in Tables 7.12–7.14. The physical and geometric parameters concerned with these plates have been taken to be the same as stated in Section 7.1. The ratio of Young's moduli (E_r) and the ratio of mass densities (ρ_r) of the FG material constituents are assumed to be 2.0 and

Table 7.6 Effect of k on the first six non-dimensional frequencies of C-C-C FG right-angled triangle (Fig. 7.2(a)) for $\theta = 0$, $E_r = 2.0$, and $\rho_r = 1.0$

μ	k	λ_1	λ_2	λ_3	λ_4	λ_5	λ_6
1.0	0.1	91.0472	154.8207	190.7053	241.1012	290.5188	344.4903
	0.2	88.8830	151.1406	186.1722	235.3702	283.6130	336.3016
	1.0	81.2413	138.1464	170.1662	215.1344	259.2296	307.3883
	2.0	78.4866	133.4621	164.3961	207.8396	250.4396	296.9653
1.5	0.1	63.5941	103.6130	137.4716	164.7291	207.8836	247.8950
	0.2	62.0825	101.1501	134.2039	160.8134	202.9421	242.0024
	1.0	56.7450	92.4538	122.6658	146.9876	185.4943	221.1964
	2.0	54.8208	89.3189	118.5064	142.0035	179.2045	213.6961
2.0	0.1	51.8958	81.3229	116.2544	127.2009	167.5871	184.1184
	0.2	50.6622	79.3899	113.4910	124.1773	163.6035	179.7419
	1.0	46.3065	72.5644	103.7337	113.5012	149.5378	164.2887
	2.0	44.7364	70.1038	100.2163	109.6526	144.4673	158.7180
2.5	0.1	45.5936	69.5275	98.9260	106.8867	148.7133	160.8342
	0.2	44.5098	67.8748	96.5745	104.3459	145.1783	157.0111
	1.0	40.6831	62.0393	88.2716	95.3749	132.6967	143.5122
	2.0	39.3036	59.9357	85.2785	92.1409	128.1972	138.6459
3.0	0.1	41.6979	61.2381	85.8599	99.1123	139.5690	151.6590
	0.2	40.7067	59.7824	83.8190	96.7564	136.2514	148.0540
	1.0	37.2070	54.6427	76.6127	88.4378	124.5372	135.3252
	2.0	35.9453	52.7898	74.0149	85.4391	120.3144	130.7365

Table 7.7 Effect of k on the first six non-dimensional frequencies of C-C-S FG right-angled triangle (Fig. 7.2(a)) for $\theta = 0$, $E_r = 2.0$, and $\rho_r = 1.0$

μ	k	λ_1	λ_2	λ_3	λ_4	λ_5	λ_6
1.0	0.1	71.2630	129.1405	163.5442	216.1318	268.2843	314.6799
	0.2	69.5691	126.0708	159.6567	210.9943	261.9071	307.1998
	1.0	63.5879	115.2320	145.9303	192.8542	239.3898	280.7885
	2.0	61.4318	111.3246	140.9821	186.3149	231.2725	271.2675
1.5	0.1	49.5298	86.3298	117.5549	146.0365	186.6522	228.8853
	0.2	48.3524	84.2777	114.7606	142.5651	182.2154	223.4446
	1.0	44.1954	77.0320	104.8941	130.3082	166.5496	204.2341
	2.0	42.6968	74.4199	101.3373	125.8897	160.9022	197.3089
2.0	0.1	40.1110	67.3078	98.9876	107.7777	154.1249	183.8509
	0.2	39.1575	65.7079	96.6347	105.2157	150.4613	179.4807
	1.0	35.7910	60.0587	88.3266	96.1699	137.5255	164.0500
	2.0	34.5774	58.0222	85.3316	92.9089	132.8622	158.4874
2.5	0.1	34.9769	56.4548	84.9497	91.4952	139.2663	167.6156
	0.2	34.1455	55.1128	82.9304	89.3204	135.9559	163.6313
	1.0	31.2098	50.3746	75.8006	81.6411	124.2672	149.5632
	2.0	30.1516	48.6664	73.2303	78.8728	120.0535	144.4918
3.0	0.1	31.7412	49.0864	75.6319	85.3760	131.7924	160.4156
	0.2	30.9867	47.9196	73.8341	83.3466	128.6596	156.6025
	1.0	28.3226	43.7997	67.4863	76.1809	117.5982	143.1387
	2.0	27.3623	42.3145	65.1980	73.5977	113.6106	138.2851

Table 7.8 Effect of k on the first six non-dimensional frequencies of S-S-S FG right-angled triangle (Fig. 7.2(a)) for $\theta = 0$, $E_r = 2.0$, and $\rho_r = 1.0$

μ	k	λ_1	λ_2	λ_3	λ_4	λ_5	λ_6
1.0	0.1	47.9233	99.3824	128.9791	176.4169	232.3638	278.5150
	0.2	46.7842	97.0200	125.9132	172.2234	226.8404	271.8945
	1.0	42.7619	88.6788	115.0879	157.4166	207.3380	248.5186
	2.0	41.3120	85.6718	111.1855	152.0789	200.3075	240.0918
1.5	0.1	33.3722	65.1384	93.2169	125.9122	169.3949	197.5589
	0.2	32.5789	63.5900	91.0011	122.9192	165.3684	192.8628
	1.0	29.7780	58.1229	83.1773	112.3513	151.1509	176.2816
	2.0	28.7682	56.1521	80.3569	108.5417	146.0257	170.3042
2.0	0.1	27.0194	49.9471	79.9467	104.5365	139.2573	167.1121
	0.2	26.3771	48.7598	78.0464	102.0516	135.9471	163.1398
	1.0	24.1094	44.5678	71.3364	93.2778	124.2591	149.1140
	2.0	23.2919	43.0565	68.9175	90.1149	120.0457	144.0578
2.5	0.1	23.4781	42.6898	72.4929	87.0437	119.2492	145.1404
	0.2	22.9200	41.6751	70.7697	84.9746	116.4146	141.6903
	1.0	20.9495	38.0921	64.6854	77.6690	106.4059	129.5086
	2.0	20.2391	36.8005	62.4920	75.0354	102.7979	125.1172
3.0	0.1	21.2926	38.4803	67.1948	71.3749	107.6519	128.4438
	0.2	20.7864	37.5656	65.5975	69.6783	105.0930	125.3906
	1.0	18.9993	34.3359	59.9578	63.6877	96.0577	114.6103
	2.0	18.3551	33.1717	57.9248	61.5282	92.8005	110.7240

Table 7.9 Effect of k on the first six non-dimensional frequencies of C-C-F FG right-angled triangle (Fig. 7.2(a)) for $\theta = 0$, $E_r = 2.0$, and $\rho_r = 1.0$

μ	k	λ_1	λ_2	λ_3	λ_4	λ_5	λ_6
1.0	0.1	28.2419	61.9956	87.7987	114.9823	166.0811	207.7691
	0.2	27.5706	60.5219	85.7117	112.2492	162.1333	202.8303
	1.0	25.2003	55.3186	78.3427	102.5986	148.1940	185.3921
	2.0	24.3458	53.4428	75.6863	99.1197	143.1690	179.1058
1.5	0.1	19.2022	40.4770	62.9945	75.6968	115.1200	137.8951
	0.2	18.7458	39.5149	61.4971	73.8974	112.3836	134.6173
	1.0	17.1341	36.1176	56.2099	67.5441	102.7215	123.0437
	2.0	16.5531	34.8929	54.3039	65.2538	99.2384	118.8715
2.0	0.1	14.9956	29.7532	52.2692	57.3964	87.6623	96.9940
	0.2	14.6392	29.0460	51.0267	56.0320	85.5785	94.6884
	1.0	13.3806	26.5488	46.6398	51.2147	78.2210	86.5477
	2.0	12.9269	25.6485	45.0583	49.4781	75.5686	83.6130
2.5	0.1	12.5795	23.9678	42.3090	48.7249	72.5474	83.1392
	0.2	12.2805	23.3981	41.3033	47.5667	70.8229	81.1630
	1.0	11.2247	21.3864	37.7523	43.4772	64.7340	74.1851
	2.0	10.8441	20.6613	36.4722	42.0029	62.5390	71.6696
3.0	0.1	11.0219	20.4341	34.0282	44.5809	66.1853	76.3127
	0.2	10.7599	19.9483	33.2194	43.5212	64.6120	74.4987
	1.0	9.8348	18.2333	30.3633	39.7795	59.0571	68.0938
	2.0	9.5014	17.6150	29.3338	38.4306	57.0545	65.7848

Table 7.10 Effect of k on the first six non-dimensional frequencies of C-S-F FG right-angled triangle (Fig. 7.2(a)) for $\theta = 0$, $E_r = 2.0$, and $\rho_r = 1.0$

μ	k	λ_1	λ_2	λ_3	λ_4	λ_5	λ_6
1.0	0.1	17.4411	46.7136	73.4769	97.7727	150.6387	194.8939
	0.2	17.0266	45.6032	71.7303	95.4486	147.0580	190.2612
	1.0	15.5627	41.6825	65.5633	87.2425	134.4148	173.9036
	2.0	15.0350	40.2691	63.3402	84.2842	129.8570	168.0069
1.5	0.1	13.3522	33.3703	52.8577	67.2324	107.6886	137.9586
	0.2	13.0349	32.5771	51.6013	65.6343	105.1288	134.6793
	1.0	11.9142	29.7763	47.1649	59.9914	96.0904	123.1003
	2.0	11.5102	28.7666	45.5656	57.9572	92.8322	118.9262
2.0	0.1	11.1886	25.5073	44.5847	52.2049	89.3005	115.0949
	0.2	10.9226	24.9010	43.5249	50.9640	87.1778	112.3591
	1.0	9.9835	22.7601	39.7829	46.5824	79.6828	102.6991
	2.0	9.6450	21.9884	38.4339	45.0029	76.9809	99.2167
2.5	0.1	9.8524	20.6162	39.8671	45.4740	76.3031	93.2331
	0.2	9.6182	20.1262	38.9195	44.3930	74.4893	91.0170
	1.0	8.7913	18.3958	35.5734	40.5764	68.0852	83.1918
	2.0	8.4932	17.7721	34.3672	39.2005	65.7765	80.3710
3.0	0.1	8.9312	17.6209	36.4697	41.7231	65.9988	73.2754
	0.2	8.7189	17.2020	35.6028	40.7314	64.4299	71.5336
	1.0	7.9693	15.7231	32.5419	37.2295	58.8906	65.3836
	2.0	7.6990	15.1899	31.4384	35.9671	56.8938	63.1665

Table 7.11 Effect of k on the first six non-dimensional frequencies of C-F-F FG right-angled triangle (Fig. 7.2(a)) for $\theta = 0$, $E_r = 2.0$, and $\rho_r = 1.0$

μ	k	λ_1	λ_2	λ_3	λ_4	λ_5	λ_6
1.0	0.1	5.9987	22.8112	31.8762	58.0127	80.5269	112.6798
	0.2	5.8561	22.2690	31.1185	56.6337	78.6127	110.0014
	1.0	5.3527	20.3544	28.4431	51.7646	71.8540	100.5441
	2.0	5.1712	19.6642	27.4786	50.0094	69.4176	97.1348
1.5	0.1	5.6303	17.9534	29.0088	41.0118	69.1853	85.0815
	0.2	5.4965	17.5266	28.3192	40.0370	67.5408	83.0591
	1.0	5.0239	16.0198	25.8845	36.5948	61.7340	75.9182
	2.0	4.8536	15.4766	25.0068	35.3540	59.6407	73.3439
2.0	0.1	5.3557	14.9572	27.8093	31.4758	61.5841	77.7571
	0.2	5.2284	14.6017	27.1482	30.7276	60.1202	75.9088
	1.0	4.7789	13.3463	24.8142	28.0858	54.9514	69.3826
	2.0	4.6169	12.8938	23.9728	27.1335	53.0881	67.0300
2.5	0.1	5.1512	12.9362	26.7216	27.1358	56.0674	71.4806
	0.2	5.0287	12.6287	26.0864	26.4908	54.7346	69.7815
	1.0	4.5964	11.5429	23.8437	24.2133	50.0288	63.7821
	2.0	4.4405	11.1515	23.0352	23.3922	48.3325	61.6193
3.0	0.1	4.9914	11.5374	24.1492	26.5355	51.7979	56.1124
	0.2	4.8727	11.2631	23.5752	25.9048	50.5667	54.7785
	1.0	4.4538	10.2948	21.5483	23.6776	46.2193	50.0690
	2.0	4.3028	9.9457	20.8176	22.8748	44.6520	48.3712

1.0, respectively, along with the power-law indices $k = 0.1$, 0.2, 1.0, and 2.0. In these tabulations, it may be noted that the non-dimensional frequencies are decreasing with an increase in k irrespective of the boundary support and geometry considered. Other than this, the descending behavior with an increase in μ can also be indicated in Tables 7.6–7.11. After analyzing the present results, one may also find the similar effect of power-law exponents (k) on the frequencies of FG triangular plates for other BCs.

7.4.2 Effect of the Ratio of Young's Moduli (E_r)

In this part, the effect of the ratio of Young's moduli (E_r) on the first six non-dimensional frequencies has been performed in Tables 7.15–7.23. Especially, the common parameters in these computations are $k = 1.0$ and $\rho_r = 1.0$. Here, increasing values of the ratio of Young's moduli (E_r) are considered to be 1.0, 1.5, 2.0, 2.5, and 3.0. Unlike Tables 7.6–7.11, FG right-angled triangular plates (Fig. 7.2(a)) subjected to samecombination of

Table 7.12 Effect of k on the first six non-dimensional frequencies of equilateral FG triangular plates (Fig. 7.2(b)) with $\theta = 1/\sqrt{3}$ and $\mu = \sqrt{3}/2$

k	BCs	λ_1	λ_2	λ_3	λ_4	λ_5	λ_6
0.1	C-C-C	96.1142	183.7803	183.7803	293.5942	310.8578	310.8578
	C-C-S	79.2065	160.7861	161.0844	265.9415	287.1341	291.1617
	C-C-F	38.8604	93.1608	99.0763	171.5459	198.5175	200.1511
	S-S-S	51.0889	120.9208	120.9208	212.9728	236.0465	236.0465
	S-C-S	64.2459	139.6457	140.1539	246.0838	261.1685	265.6640
	S-C-F	25.7909	73.3945	82.4641	148.2524	177.4789	180.2259
	S-F-S	15.6200	56.4846	66.7480	121.1134	151.6160	153.6938
	F-F-F	34.2342	35.5200	35.5200	108.6637	108.6637	115.2961
	F-C-F	8.6622	34.3530	37.4139	92.8219	96.3350	112.0160
	F-F-S	22.0114	26.0639	70.0937	73.3966	90.1712	185.3062
0.2	C-C-C	93.8295	179.4117	179.4117	286.6154	303.4686	303.4686
	C-C-S	77.3238	156.9641	157.2553	259.6200	280.3089	284.2406
	C-C-F	37.9366	90.9463	96.7212	167.4682	193.7986	195.3935
	S-S-S	49.8745	118.0464	118.0464	207.9103	230.4355	230.4355
	S-C-S	62.7188	136.3262	136.8223	240.2343	254.9605	259.3490
	S-C-F	25.1778	71.6498	80.5039	144.7283	173.2602	175.9419
	S-F-S	15.2487	55.1419	65.1614	118.2345	148.0120	150.0405
	F-F-F	33.4205	34.6757	34.6757	106.0807	106.0807	112.5554
	F-C-F	8.4563	33.5364	36.5246	90.6155	94.0451	109.3533
	F-F-S	21.4882	25.4444	68.4276	71.6519	88.0278	180.9014
1.0	C-C-C	85.7626	163.9869	163.9869	261.9738	277.3781	277.3781
	C-C-S	70.6759	143.4692	143.7354	237.2993	256.2095	259.8033
	C-C-F	34.6751	83.1273	88.4057	153.0702	177.1369	178.5946
	S-S-S	45.5866	107.8975	107.8975	190.0354	210.6240	210.6240
	S-C-S	57.3266	124.6057	125.0591	219.5804	233.0404	237.0517
	S-C-F	23.0132	65.4898	73.5826	132.2854	158.3643	160.8154
	S-F-S	13.9377	50.4011	59.5592	108.0694	135.2868	137.1409
	F-F-F	30.5472	31.6945	31.6945	96.9605	96.9605	102.8786
	F-C-F	7.7293	30.6531	33.3844	82.8249	85.9596	99.9518
	F-F-S	19.6407	23.2568	62.5446	65.4917	80.4597	165.3485
2.0	C-C-C	82.8545	158.4264	158.4264	253.0908	267.9727	267.9727
	C-C-S	68.2794	138.6044	138.8616	229.2529	247.5219	250.9938
	C-C-F	33.4993	80.3086	85.4080	147.8799	171.1305	172.5388
	S-S-S	44.0408	104.2389	104.2389	183.5916	203.4821	203.4821
	S-C-S	55.3827	120.3805	120.8186	212.1348	225.1384	229.0137
	S-C-F	22.2328	63.2692	71.0875	127.7999	152.9944	155.3624
	S-F-S	13.4651	48.6921	57.5396	104.4049	130.6994	132.4907
	F-F-F	29.5114	30.6198	30.6198	93.6728	93.6728	99.3901
	F-C-F	7.4672	29.6137	32.2524	80.0164	83.0449	96.5626
	F-F-S	18.9748	22.4682	60.4238	63.2710	77.7314	159.7418

Continued

Table 7.12 Effect of k on the first six non-dimensional frequencies of equilateral FG triangular plates (Fig. 7.2(b)) with $\theta = 1/\sqrt{3}$ and $\mu = \sqrt{3}/2$—cont'd

k	BCs	λ_1	λ_2	λ_3	λ_4	λ_5	λ_6
5.0	C-C-C	79.4007	151.8224	151.8224	242.5406	256.8022	256.8022
	C-C-S	65.4332	132.8267	133.0732	219.6965	237.2039	240.5311
	C-C-F	32.1029	76.9609	81.8477	141.7155	163.9969	165.3465
	S-S-S	42.2050	99.8936	99.8936	175.9386	194.9999	194.9999
	S-C-S	53.0741	115.3625	115.7823	203.2919	215.7535	219.4672
	S-C-F	21.3061	60.6318	68.1242	122.4725	146.6168	148.8861
	S-F-S	12.9038	46.6624	55.1411	100.0528	125.2512	126.9678
	F-F-F	28.2812	29.3434	29.3434	89.7680	89.7680	95.2470
	F-C-F	7.1559	28.3793	30.9080	76.6809	79.5831	92.5373
	F-F-S	18.1838	21.5316	57.9050	60.6335	74.4912	153.0830

Table 7.13 Effect of k on the first six non-dimensional frequencies of FG right-angled triangular plates (Fig. 7.2(c)) with angles 30°, 60°, and 90°

k	BCs	λ_1	λ_2	λ_3	λ_4	λ_5	λ_6
0.1	C-C-C	171.4750	273.4001	378.0292	436.6760	557.5210	638.2764
	C-C-S	133.0728	227.5203	322.4896	376.8116	504.7479	608.4365
	C-C-F	50.7597	103.8812	172.2018	197.2515	303.4861	338.8338
	S-S-S	89.7121	169.7380	257.5380	344.1051	462.1235	540.3853
	S-C-S	116.0340	206.8043	296.1916	365.4608	507.7390	596.6455
	S-C-F	36.6036	87.7666	144.9766	176.0460	294.0350	376.8694
	S-F-S	22.3134	64.3916	126.2444	139.7782	279.1071	346.0023
	F-F-F	24.8248	52.9835	78.6936	98.3006	139.4496	225.1636
	F-C-F	16.4769	49.2063	85.1041	107.1639	196.0819	242.0130
	F-F-S	31.1018	67.2067	86.4201	165.3371	217.8876	234.3610
0.2	C-C-C	167.3990	266.9013	369.0433	426.2960	544.2685	623.1044
	C-C-S	129.9096	222.1121	314.8239	367.8546	492.7498	593.9737
	C-C-F	49.5532	101.4119	168.1085	192.5628	296.2721	330.7796
	S-S-S	87.5796	165.7033	251.4162	335.9256	451.1386	527.5401
	S-C-S	113.2758	201.8885	289.1510	356.7737	495.6699	582.4630
	S-C-F	35.7335	85.6803	141.5304	171.8613	287.0457	367.9110
	S-F-S	21.7830	62.8609	123.2435	136.4556	272.4726	337.7777
	F-F-F	24.2347	51.7241	76.8231	95.9639	136.1349	219.8114
	F-C-F	16.0853	48.0367	83.0812	104.6166	191.4209	236.2602
	F-F-S	30.3625	65.6092	84.3659	161.4070	212.7083	228.7901
1.0	C-C-C	153.0070	243.9546	337.3150	389.6455	497.4754	569.5334
	C-C-S	118.7408	203.0162	287.7571	336.2286	450.3860	542.9073
	C-C-F	45.2929	92.6931	153.6555	176.0073	270.8003	302.3411
	S-S-S	80.0500	151.4571	229.8009	307.0447	412.3523	482.1852
	S-C-S	103.5370	184.5313	264.2914	326.1003	453.0550	532.3862

Table 7.13 Effect of *k* on the first six non-dimensional frequencies of FG right-angled triangular plates (Fig. 7.2(c)) with angles 30°, 60°, and 90°—cont'd

k	BCs	λ_1	λ_2	λ_3	λ_4	λ_5	λ_6
	S-C-F	32.6613	78.3140	129.3624	157.0856	262.3671	336.2801
	S-F-S	19.9102	57.4565	112.6477	124.7239	249.0470	308.7375
	F-F-F	22.1511	47.2771	70.2183	87.7135	124.4308	200.9133
	F-C-F	14.7024	43.9067	75.9383	95.6222	174.9637	215.9479
	F-F-S	27.7521	59.9685	77.1126	147.5301	194.4209	209.1200
2.0	C-C-C	147.8188	235.6826	325.8773	376.4334	480.6069	550.2216
	C-C-S	114.7145	196.1323	277.9998	324.8277	435.1142	524.4983
	C-C-F	43.7571	89.5500	148.4453	170.0392	261.6180	292.0892
	S-S-S	77.3356	146.3214	222.0088	296.6333	398.3702	465.8352
	S-C-S	100.0263	178.2742	255.3298	315.0428	437.6927	514.3339
	S-C-F	31.5538	75.6585	124.9760	151.7591	253.4707	324.8775
	S-F-S	19.2351	55.5083	108.8280	120.4948	240.6022	298.2687
	F-F-F	21.4000	45.6740	67.8373	84.7393	120.2115	194.1007
	F-C-F	14.2038	42.4179	73.3634	92.3799	169.0310	208.6255
	F-F-S	26.8111	57.9351	74.4978	142.5277	187.8284	202.0292
5.0	C-C-C	141.6570	225.8581	312.2930	360.7417	460.5727	527.2855
	C-C-S	109.9326	187.9565	266.4113	311.2872	416.9764	502.6345
	C-C-F	41.9331	85.8171	142.2573	162.9511	250.7124	279.9135
	S-S-S	74.1119	140.2220	212.7543	284.2681	381.7641	446.4168
	S-C-S	95.8566	170.8428	244.6863	301.9102	419.4475	492.8938
	S-C-F	30.2385	72.5047	119.7664	145.4330	242.9048	311.3349
	S-F-S	18.4333	53.1944	104.2915	115.4719	230.5727	285.8354
	F-F-F	20.5080	43.7701	65.0095	81.2069	115.2005	186.0096
	F-C-F	13.6117	40.6497	70.3052	88.5290	161.9849	199.9289
	F-F-S	25.6934	55.5200	71.3924	136.5864	179.9987	193.6075

Table 7.14 Effect of *k* on the first six non-dimensional frequencies of isosceles FG triangular plates (Fig. 7.2(d)) with angles 30°, 30°, and 120°

k	BCs	λ_1	λ_2	λ_3	λ_4	λ_5	λ_6
0.1	C-C-C	137.6361	213.6173	296.0524	354.9282	453.3632	616.0397
	C-C-S	104.1299	172.6731	261.8057	313.4529	425.4168	547.3025
	C-C-F	33.8016	68.5302	111.8262	154.5037	243.4675	320.7225
	S-S-S	71.6489	135.0420	214.8276	256.3671	386.7775	508.6849
	S-C-S	86.6240	152.3939	239.0463	288.2326	401.4156	526.6853
	S-C-F	18.6026	47.2619	81.2986	131.6930	209.0212	253.5735
	S-F-S	23.9881	59.8452	106.6091	150.9550	208.1678	317.1308
	F-F-F	13.3306	27.4809	49.3394	54.9804	99.5320	138.8216
	F-C-F	5.5679	21.1764	37.5645	55.5857	92.0183	138.1714
	F-F-S	22.1655	43.6699	88.7802	93.6692	170.7119	202.5393

Continued

Table 7.14 Effect of *k* on the first six non-dimensional frequencies of isosceles FG triangular plates (Fig. 7.2(d)) with angles 30°, 30°, and 120°—cont'd

k	BCs	λ_1	λ_2	λ_3	λ_4	λ_5	λ_6
0.2	C-C-C	134.3645	208.5395	289.0151	346.4914	442.5866	601.3962
	C-C-S	101.6547	168.5686	255.5825	306.0020	415.3045	534.2929
	C-C-F	32.9981	66.9012	109.1680	150.8311	237.6802	313.0988
	S-S-S	69.9458	131.8320	209.7210	250.2731	377.5837	496.5933
	S-C-S	84.5649	148.7714	233.3641	281.3812	391.8738	514.1658
	S-C-F	18.1604	46.1385	79.3661	128.5626	204.0527	247.5460
	S-F-S	23.4179	58.4227	104.0750	147.3667	203.2196	309.5924
	F-F-F	13.0137	26.8277	48.1665	53.6735	97.1661	135.5217
	F-C-F	5.4355	20.6731	36.6716	54.2644	89.8310	134.8870
	F-F-S	21.6386	42.6319	86.6699	91.4427	166.6540	197.7249
1.0	C-C-C	122.8126	190.6105	264.1672	316.7021	404.5355	549.6916
	C-C-S	92.9150	154.0760	233.6090	279.6937	379.5990	488.3574
	C-C-F	30.1611	61.1494	99.7824	137.8635	217.2458	286.1804
	S-S-S	63.9323	120.4979	191.6904	228.7561	345.1212	453.8990
	S-C-S	77.2945	135.9809	213.3008	257.1897	358.1827	469.9607
	S-C-F	16.5990	42.1717	72.5427	117.5095	186.5094	226.2634
	S-F-S	21.4046	53.3998	95.1272	134.6970	185.7480	282.9755
	F-F-F	11.8949	24.5212	44.0255	49.0590	88.8123	123.8704
	F-C-F	4.9682	18.8957	33.5188	49.5990	82.1078	123.2902
	F-F-S	19.7783	38.9666	79.2185	83.5810	152.3261	180.7256
2.0	C-C-C	118.6482	184.1472	255.2098	305.9633	390.8184	531.0526
	C-C-S	89.7644	148.8516	225.6877	270.2098	366.7275	471.7981
	C-C-F	29.1384	59.0760	96.3989	133.1888	209.8794	276.4765
	S-S-S	61.7644	116.4120	185.1906	220.9994	333.4188	438.5081
	S-C-S	74.6735	131.3701	206.0681	248.4689	346.0374	454.0252
	S-C-F	16.0362	40.7418	70.0829	113.5250	180.1852	218.5912
	S-F-S	20.6788	51.5892	91.9016	130.1296	179.4496	273.3803
	F-F-F	11.4916	23.6897	42.5326	47.3955	85.8008	119.6701
	F-C-F	4.7997	18.2550	32.3822	47.9172	79.3237	119.1096
	F-F-S	19.1076	37.6454	76.5323	80.7469	147.1610	174.5975
5.0	C-C-C	113.7024	176.4710	244.5713	293.2091	374.5271	508.9156
	C-C-S	86.0226	142.6467	216.2799	258.9460	351.4404	452.1311
	C-C-F	27.9237	56.6134	92.3805	127.6368	201.1305	264.9516
	S-S-S	59.1898	111.5593	177.4709	211.7870	319.5202	420.2288
	S-C-S	71.5608	125.8939	197.4781	238.1114	331.6127	435.0991
	S-C-F	15.3677	39.0434	67.1615	108.7927	172.6742	209.4792
	S-F-S	19.8168	49.4386	88.0707	124.7051	171.9692	261.9844
	F-F-F	11.0125	22.7022	40.7597	45.4198	82.2242	114.6817
	F-C-F	4.5997	17.4940	31.0324	45.9198	76.0171	114.1445
	F-F-S	18.3111	36.0761	73.3421	77.3809	141.0265	167.3194

Table 7.15 Effect of E_r on the first six non-dimensional frequencies of C-C-C FG right-angled triangle (Fig. 7.2(a)) for $\theta = 0$, $k = 1.0$, and $\rho_r = 1.0$

μ	E_r	λ_1	λ_2	λ_3	λ_4	λ_5	λ_6
1.0	1.0	93.8094	159.5177	196.4910	248.4158	299.3326	354.9414
	2.0	81.2413	138.1464	170.1662	215.1344	259.2296	307.3883
	2.5	78.4866	133.4621	164.3961	207.8396	250.4396	296.9653
	3.0	76.5950	130.2457	160.4342	202.8306	244.4040	289.8085
1.5	1.0	65.5234	106.7565	141.6423	169.7267	214.1904	255.4156
	2.0	56.7450	92.4538	122.6658	146.9876	185.4943	221.1964
	2.5	54.8208	89.3189	118.5064	142.0035	179.2045	213.6961
	3.0	53.4997	87.1663	115.6504	138.5812	174.8857	208.5460
2.0	1.0	53.4702	83.7901	119.7814	131.0599	172.6714	189.7042
	2.0	46.3065	72.5644	103.7337	113.5012	149.5378	164.2887
	2.5	44.7364	70.1038	100.2163	109.6526	144.4673	158.7180
	3.0	43.6582	68.4144	97.8011	107.0100	140.9856	154.8929
2.5	1.0	46.9768	71.6368	101.9272	110.1294	153.2250	165.7136
	2.0	40.6831	62.0393	88.2716	95.3749	132.6967	143.5122
	2.5	39.3036	59.9357	85.2785	92.1409	128.1972	138.6459
	3.0	38.3564	58.4912	83.2232	89.9203	125.1077	135.3046
3.0	1.0	42.9629	63.0959	88.4648	102.1192	143.8032	156.2600
	2.0	37.2070	54.6427	76.6127	88.4378	124.5372	135.3252
	2.5	35.9453	52.7898	74.0149	85.4391	120.3144	130.7365
	3.0	35.0791	51.5176	72.2312	83.3800	117.4148	127.5858

Table 7.16 Effect of E_r on the first six non-dimensional frequencies of C-C-S FG right-angled triangle (Fig. 7.2(a)) for $\theta = 0$, $k = 1.0$, and $\rho_r = 1.0$

μ	E_r	λ_1	λ_2	λ_3	λ_4	λ_5	λ_6
1.0	1.0	73.4250	133.0584	168.5058	222.6888	276.4235	324.2267
	2.0	63.5879	115.2320	145.9303	192.8542	239.3898	280.7885
	2.5	61.4318	111.3246	140.9821	186.3149	231.2725	271.2675
	3.0	59.9513	108.6417	137.5844	181.8247	225.6989	264.7300
1.5	1.0	51.0324	88.9488	121.1213	150.4670	192.3149	235.8292
	2.0	44.1954	77.0320	104.8941	130.3082	166.5496	204.2341
	2.5	42.6968	74.4199	101.3373	125.8897	160.9022	197.3089
	3.0	41.6678	72.6264	98.8951	122.8558	157.0245	192.5538
2.0	1.0	41.3279	69.3498	101.9907	111.0474	158.8008	189.4286
	2.0	35.7910	60.0587	88.3266	96.1699	137.5255	164.0500

Continued

Table 7.16 Effect of E_r on the first six non-dimensional frequencies of C-C-S FG right-angled triangle (Fig. 7.2(a)) for $\theta = 0$, $k = 1.0$, and $\rho_r = 1.0$—cont'd

μ	E_r	λ_1	λ_2	λ_3	λ_4	λ_5	λ_6
	2.5	34.5774	58.0222	85.3316	92.9089	132.8622	158.4874
	3.0	33.7441	56.6239	83.2751	90.6698	129.6603	154.6678
2.5	1.0	36.0380	58.1675	87.5269	94.2710	143.4914	172.7007
	2.0	31.2098	50.3746	75.8006	81.6411	124.2672	149.5632
	2.5	30.1516	48.6664	73.2303	78.8728	120.0535	144.4918
	3.0	29.4249	47.4936	71.4655	76.9720	117.1603	141.0095
3.0	1.0	32.7042	50.5756	77.9265	87.9661	135.7907	165.2824
	2.0	28.3226	43.7997	67.4863	76.1809	117.5982	143.1387
	2.5	27.3623	42.3145	65.1980	73.5977	113.6106	138.2851
	3.0	26.7028	41.2948	63.6267	71.8240	110.8726	134.9525

Table 7.17 Effect of E_r on the first six non-dimensional frequencies of S-S-S FG right-angled triangle (Fig. 7.2(a)) for $\theta = 0$, $k = 1.0$, and $\rho_r = 1.0$

μ	E_r	λ_1	λ_2	λ_3	λ_4	λ_5	λ_6
1.0	1.0	49.3772	102.3974	132.8920	181.7691	239.4133	286.9646
	2.0	42.7619	88.6788	115.0879	157.4166	207.3380	248.5186
	2.5	41.3120	85.6718	111.1855	152.0789	200.3075	240.0918
	3.0	40.3164	83.6072	108.5059	148.4138	195.4801	234.3056
1.5	1.0	34.3846	67.1145	96.0449	129.7321	174.5341	203.5525
	2.0	29.7780	58.1229	83.1773	112.3513	151.1509	176.2816
	2.5	28.7682	56.1521	80.3569	108.5417	146.0257	170.3042
	3.0	28.0749	54.7988	78.4204	105.9258	142.5065	166.1999
2.0	1.0	27.8391	51.4624	82.3722	107.7080	143.4821	172.1820
	2.0	24.1094	44.5678	71.3364	93.2778	124.2591	149.1140
	2.5	23.2919	43.0565	68.9175	90.1149	120.0457	144.0578
	3.0	22.7305	42.0189	67.2566	87.9432	117.1526	140.5860
2.5	1.0	24.1904	43.9850	74.6922	89.6844	122.8670	149.5437
	2.0	20.9495	38.0921	64.6854	77.6690	106.4059	129.5086
	2.5	20.2391	36.8005	62.4920	75.0354	102.7979	125.1172
	3.0	19.7514	35.9136	60.9859	73.2270	100.3205	122.1019
3.0	1.0	21.9385	39.6477	69.2333	73.5403	110.9179	132.3405
	2.0	18.9993	34.3359	59.9578	63.6877	96.0577	114.6103
	2.5	18.3551	33.1717	57.9248	61.5282	92.8005	110.7240
	3.0	17.9127	32.3722	56.5288	60.0454	90.5640	108.0556

Table 7.18 Effect of E_r on the first six non-dimensional frequencies of C-C-F FG right-angled triangle (Fig. 7.2(a)) for $\theta = 0$, $k = 1.0$, and $\rho_r = 1.0$

μ	E_r	λ_1	λ_2	λ_3	λ_4	λ_5	λ_6
1.0	1.0	29.0988	63.8764	90.4624	118.4707	171.1197	214.0724
	2.0	25.2003	55.3186	78.3427	102.5986	148.1940	185.3921
	2.5	24.3458	53.4428	75.6863	99.1197	143.1690	179.1058
	3.0	23.7590	52.1549	73.8622	96.7309	139.7187	174.7894
1.5	1.0	19.7848	41.7050	64.9056	77.9933	118.6126	142.0786
	2.0	17.1341	36.1176	56.2099	67.5441	102.7215	123.0437
	2.5	16.5531	34.8929	54.3039	65.2538	99.2384	118.8715
	3.0	16.1542	34.0520	52.9952	63.6812	96.8468	116.0067
2.0	1.0	15.1506	30.6559	53.8549	59.1377	90.3218	99.9366
	2.0	13.3806	26.5488	46.6398	51.2147	78.2210	86.5477
	2.5	12.9269	25.6485	45.0583	49.4781	75.5686	83.6130
	3.0	12.6153	25.0304	43.9724	48.2857	73.7474	81.5979
2.5	1.0	12.9612	24.6949	43.5926	50.2031	74.7484	85.6615
	2.0	11.2247	21.3864	37.7523	43.4772	64.7340	74.1851
	2.5	10.8441	20.6613	36.4722	42.0029	62.5390	71.6696
	3.0	10.5828	20.1633	35.5932	40.9907	61.0318	69.9423
3.0	1.0	11.3563	21.0540	35.0606	45.9334	68.1932	78.6279
	2.0	9.8348	18.2333	30.3633	39.7795	59.0571	68.0938
	2.5	9.5014	17.6150	29.3338	38.4306	57.0545	65.7848
	3.0	9.2724	17.1905	28.6268	37.5044	55.6795	64.1994

Table 7.19 Effect of E_r on the first six non-dimensional frequencies of C-S-F FG right-angled triangle (Fig. 7.2(a)) for $\theta = 0$, $k = 1.0$, and $\rho_r = 1.0$

μ	E_r	λ_1	λ_2	λ_3	λ_4	λ_5	λ_6
1.0	1.0	17.9703	48.1308	75.7060	100.7389	155.2088	200.8066
	2.0	15.5627	41.6825	65.5633	87.2425	134.4148	173.9036
	2.5	15.0350	40.2691	63.3402	84.2842	129.8570	168.0069
	3.0	14.6727	39.2986	61.8137	82.2530	126.7274	163.9579
1.5	1.0	13.7573	34.3827	54.4613	69.2721	110.9556	142.1440
	2.0	11.9142	29.7763	47.1649	59.9914	96.0904	123.1003
	2.5	11.5102	28.7666	45.5656	57.9572	92.8322	118.9262
	3.0	11.2328	28.0733	44.4675	56.5604	90.5949	116.0601
2.0	1.0	11.5280	26.2811	45.9373	53.7887	92.0098	118.5867
	2.0	9.9835	22.7601	39.7829	46.5824	79.6828	102.6991

Continued

Table 7.19 Effect of E_r on the first six non-dimensional frequencies of C-S-F FG right-angled triangle (Fig. 7.2(a)) for $\theta = 0$, $k = 1.0$, and $\rho_r = 1.0$—cont'd

μ	E_r	λ_1	λ_2	λ_3	λ_4	λ_5	λ_6
	2.5	9.6450	21.9884	38.4339	45.0029	76.9809	99.2167
	3.0	9.4126	21.4584	37.5077	43.9183	75.1256	96.8256
2.5	1.0	10.1513	21.2417	41.0766	46.8536	78.6180	96.0617
	2.0	8.7913	18.3958	35.5734	40.5764	68.0852	83.1918
	2.5	8.4932	17.7721	34.3672	39.2005	65.7765	80.3710
	3.0	8.2885	17.3438	33.5389	38.2558	64.1913	78.4340
3.0	1.0	9.2021	18.1554	37.5761	42.9889	68.0010	75.4984
	2.0	7.9693	15.7231	32.5419	37.2295	58.8906	65.3836
	2.5	7.6990	15.1899	31.4384	35.9671	56.8938	63.1665
	3.0	7.5135	14.8239	30.6808	35.1003	55.5226	61.6442

Table 7.20 Effect of E_r on the first six non-dimensional frequencies of C-F-F FG right-angled triangle (Fig. 7.2(a)) for $\theta = 0$, $k = 1.0$, and $\rho_r = 1.0$

μ	E_r	λ_1	λ_2	λ_3	λ_4	λ_5	λ_6
1.0	1.0	6.1807	23.5033	32.8433	59.7726	82.9699	116.0983
	2.0	5.3527	20.3544	28.4431	51.7646	71.8540	100.5441
	2.5	5.1712	19.6642	27.4786	50.0094	69.4176	97.1348
	3.0	5.0465	19.1903	26.8164	48.8042	67.7446	94.7939
1.5	1.0	5.8011	18.4980	29.8889	42.2561	71.2843	87.6627
	2.0	5.0239	16.0198	25.8845	36.5948	61.7340	75.9182
	2.5	4.8536	15.4766	25.0068	35.3540	59.6407	73.3439
	3.0	4.7366	15.1036	24.4041	34.5019	58.2034	71.5763
2.0	1.0	5.5182	15.4110	28.6530	32.4307	63.4524	80.1162
	2.0	4.7789	13.3463	24.8142	28.0858	54.9514	69.3826
	2.5	4.6169	12.8938	23.9728	27.1335	53.0881	67.0300
	3.0	4.5056	12.5830	23.3950	26.4796	51.8087	65.4146
2.5	1.0	5.3074	13.3286	27.5323	27.9591	57.7683	73.6492
	2.0	4.5964	11.5429	23.8437	24.2133	50.0288	63.7821
	2.5	4.4405	11.1515	23.0352	23.3922	48.3325	61.6193
	3.0	4.3335	10.8828	22.4800	22.8285	47.1676	60.1343
3.0	1.0	5.1428	11.8874	24.8818	27.3406	53.3694	57.8147
	2.0	4.4538	10.2948	21.5483	23.6776	46.2193	50.0690
	2.5	4.3028	9.9457	20.8176	22.8748	44.6520	48.3712
	3.0	4.1991	9.7060	20.3159	22.3235	43.5759	47.2055

Table 7.21 Effect of E_r on the first six non-dimensional frequencies of equilateral FG triangular plates (Fig. 7.2(b)) with $\theta = 1/\sqrt{3}$ and $\mu = \sqrt{3}/2$

E_r	BCs	λ_1	λ_2	λ_3	λ_4	λ_5	λ_6
1.0	C-C-C	99.0301	189.3558	189.3558	302.5013	320.2886	320.2886
	C-C-S	81.6095	165.6640	165.9714	274.0097	295.8453	299.9950
	C-C-F	40.0393	95.9871	102.0821	176.7503	204.5401	206.2233
	S-S-S	52.6389	124.5893	124.5893	219.4339	243.2077	243.2077
	S-C-S	66.1950	143.8823	144.4059	253.5495	269.0919	273.7237
	S-C-F	26.5733	75.6211	84.9659	152.7500	182.8633	185.693
	S-F-S	16.0939	58.1982	68.7730	124.7878	156.2157	158.3566
	F-F-F	35.2728	36.5976	36.5976	111.9604	111.9604	118.7939
	F-C-F	8.9250	35.3952	38.5490	95.6379	99.2576	115.4143
	F-F-S	22.6792	26.8547	72.2202	75.6233	92.9068	190.9280
2.0	C-C-C	85.7626	163.9869	163.9869	261.9738	277.3781	277.3781
	C-C-S	70.6759	143.4692	143.7354	237.2993	256.2095	259.8033
	C-C-F	34.6751	83.1273	88.4057	153.0702	177.1369	178.5946
	S-S-S	45.5866	107.8975	107.8975	190.0354	210.6240	210.6240
	S-C-S	57.3266	124.6057	125.0591	219.5804	233.0404	237.0517
	S-C-F	23.0132	65.4898	73.5826	132.2854	158.3643	160.8154
	S-F-S	13.9377	50.4011	59.5592	108.0694	135.2868	137.1409
	F-F-F	30.5472	31.6945	31.6945	96.9605	96.9605	102.8786
	F-C-F	7.7293	30.6531	33.3844	82.8249	85.9596	99.9518
	F-F-S	19.6407	23.2568	62.5446	65.4917	80.4597	165.3485
2.5	C-C-C	82.8545	158.4264	158.4264	253.0908	267.9727	267.9727
	C-C-S	68.2794	138.6044	138.8616	229.2529	247.5219	250.9938
	C-C-F	33.4993	80.3086	85.4080	147.8799	171.1305	172.5388
	S-S-S	44.0408	104.2389	104.2389	183.5916	203.4821	203.4821
	S-C-S	55.3827	120.3805	120.8186	212.1348	225.1384	229.0137
	S-C-F	22.2328	63.2692	71.0875	127.7999	152.9944	155.3624
	S-F-S	13.4651	48.6921	57.5396	104.4049	130.6994	132.4907
	F-F-F	29.5114	30.6198	30.6198	93.6728	93.6728	99.3901
	F-C-F	7.4672	29.6137	32.2524	80.0164	83.0449	96.5626
	F-F-S	18.9748	22.4682	60.4238	63.2710	77.7314	159.7418
3.0	C-C-C	80.8577	154.6084	154.6084	246.9913	261.5146	261.5146
	C-C-S	66.6339	135.2641	135.5151	223.7280	241.5566	244.9449
	C-C-F	32.6920	78.3731	83.3497	144.3160	167.0063	168.3806
	S-S-S	42.9794	101.7267	101.7267	179.1671	198.5782	198.5782
	S-C-S	54.0480	117.4794	117.9069	207.0223	219.7126	223.4945
	S-C-F	21.6970	61.7444	69.3743	124.7199	149.3073	151.6182
	S-F-S	13.1406	47.5186	56.1529	101.8888	127.5496	129.2976
	F-F-F	28.8001	29.8818	29.8818	91.4153	91.4153	96.9948
	F-C-F	7.2872	28.9000	31.4751	78.0880	81.0435	94.2354
	F-F-S	18.5175	21.9267	58.9676	61.7462	75.8581	155.8921

Table 7.22 Effect of E_r on the first six non-dimensional frequencies of FG right-angled triangular plates (Fig. 7.2(c)) with angles 30°, 60°, and 90°

E_r	BCs	λ_1	λ_2	λ_3	λ_4	λ_5	λ_6
1.0	C-C-C	176.6773	281.6946	389.4979	449.9239	574.4351	657.6405
	C-C-S	137.1100	234.4229	332.2733	388.2433	520.0610	626.8953
	C-C-F	52.2997	107.0327	177.4260	203.2358	312.6933	349.1134
	S-S-S	92.4337	174.8875	265.3512	354.5446	476.1434	556.7795
	S-C-S	119.5542	213.0784	305.1774	376.5482	523.1429	614.7466
	S-C-F	37.7140	90.4292	149.3749	181.3869	302.9555	388.3029
	S-F-S	22.9903	66.3451	130.0744	144.0188	287.5747	356.4993
	F-F-F	25.5779	54.5909	81.0811	101.2828	143.6803	231.9947
	F-C-F	16.9768	50.6991	87.6860	110.4150	202.0306	249.3552
	F-F-S	32.0453	69.2456	89.0420	170.3531	224.4979	241.4710
2.0	C-C-C	153.0070	243.9546	337.3150	389.6455	497.4754	569.5334
	C-C-S	118.7408	203.0162	287.7571	336.2286	450.3860	542.9073
	C-C-F	45.2929	92.6931	153.6555	176.0073	270.8003	302.3411
	S-S-S	80.0500	151.4571	229.8009	307.0447	412.3523	482.1852
	S-C-S	103.5370	184.5313	264.2914	326.1003	453.0550	532.3862
	S-C-F	32.6613	78.3140	129.3624	157.0856	262.3671	336.2801
	S-F-S	19.9102	57.4565	112.6477	124.7239	249.0470	308.7375
	F-F-F	22.1511	47.2771	70.2183	87.7135	124.4308	200.9133
	F-C-F	14.7024	43.9067	75.9383	95.6222	174.9637	215.9479
	F-F-S	27.7521	59.9685	77.1126	147.5301	194.4209	209.1200
2.5	C-C-C	147.8188	235.6826	325.8773	376.4334	480.6069	550.2216
	C-C-S	114.7145	196.1323	277.9998	324.8277	435.1142	524.4983
	C-C-F	43.7571	89.5500	148.4453	170.0392	261.6180	292.0892
	S-S-S	77.3356	146.3214	222.0088	296.6333	398.3702	465.8352
	S-C-S	100.0263	178.2742	255.3298	315.0428	437.6927	514.3339
	S-C-F	31.5538	75.6585	124.9760	151.7591	253.4707	324.8775
	S-F-S	19.2351	55.5083	108.8280	120.4948	240.6022	298.2687
	F-F-F	21.4000	45.6740	67.8373	84.7393	120.2115	194.1007
	F-C-F	14.2038	42.4179	73.3634	92.3799	169.0310	208.6255
	F-F-S	26.8111	57.9351	74.4978	142.5277	187.8284	202.0292
3.0	C-C-C	144.2564	230.0026	318.0237	367.3613	469.0243	536.9613
	C-C-S	111.9499	191.4055	271.3000	316.9994	424.6280	511.8579
	C-C-F	42.7025	87.3919	144.8677	165.9413	255.3130	285.0499
	S-S-S	75.4718	142.7951	216.6584	289.4845	388.7695	454.6086
	S-C-S	97.6156	173.9778	249.1763	307.4503	427.1444	501.9385
	S-C-F	30.7934	73.8352	121.9641	148.1018	247.3621	317.0480
	S-F-S	18.7715	54.1705	106.2053	117.5909	234.8037	291.0805
	F-F-F	20.8843	44.5733	66.2024	82.6971	117.3144	189.4229
	F-C-F	13.8615	41.3957	71.5953	90.1535	164.9573	203.5977
	F-F-S	26.1649	56.5388	72.7025	139.0928	183.3018	197.1603

Table 7.23 Effect of E_r on the first six non-dimensional frequencies of isosceles FG triangular plates (Fig. 7.2(d)) with angles 30°, 30°, and 120°

E_r	BCs	λ_1	λ_2	λ_3	λ_4	λ_5	λ_6
1.0	C-C-C	141.8117	220.0980	305.0341	365.6960	467.1174	634.7292
	C-C-S	107.2890	177.9117	269.7484	322.9624	438.3232	563.9066
	C-C-F	34.8270	70.6093	115.2188	159.1910	250.8538	330.4526
	S-S-S	73.8226	139.1389	221.3450	264.1448	398.5116	524.1175
	S-C-S	89.2520	157.0172	246.2985	296.9771	413.5938	542.6639
	S-C-F	19.1669	48.6957	83.7651	135.6883	215.3625	261.2665
	S-F-S	24.7159	61.6608	109.8434	155.5346	214.4833	326.7519
	F-F-F	13.7350	28.3146	50.8362	56.6484	102.5516	143.0332
	F-C-F	5.7368	21.8189	38.7042	57.2720	94.8100	142.3632
	F-F-S	22.8380	44.9948	91.4736	96.5110	175.8910	208.6840
2.0	C-C-C	122.8126	190.6105	264.1672	316.7021	404.5355	549.6916
	C-C-S	92.9150	154.0760	233.6090	279.6937	379.5990	488.3574
	C-C-F	30.1611	61.1494	99.7824	137.8635	217.2458	286.1804
	S-S-S	63.9323	120.4979	191.6904	228.7561	345.1212	453.8990
	S-C-S	77.2945	135.9809	213.3008	257.1897	358.1827	469.9607
	S-C-F	16.5990	42.1717	72.5427	117.5095	186.5094	226.2634
	S-F-S	21.4046	53.3998	95.1272	134.6970	185.7480	282.9755
	F-F-F	11.8949	24.5212	44.0255	49.0590	88.8123	123.8704
	F-C-F	4.9682	18.8957	33.5188	49.5990	82.1078	123.2902
	F-F-S	19.7783	38.9666	79.2185	83.5810	152.3261	180.7256
2.5	C-C-C	118.6482	184.1472	255.2098	305.9633	390.8184	531.0526
	C-C-S	89.7644	148.8516	225.6877	270.2098	366.7275	471.7981
	C-C-F	29.1384	59.0760	96.3989	133.1888	209.8794	276.4765
	S-S-S	61.7644	116.4120	185.1906	220.9994	333.4188	438.5081
	S-C-S	74.6735	131.3701	206.0681	248.4689	346.0374	454.0252
	S-C-F	16.0362	40.7418	70.0829	113.5250	180.1852	218.5912
	S-F-S	20.6788	51.5892	91.9016	130.1296	179.4496	273.3803
	F-F-F	11.4916	23.6897	42.5326	47.3955	85.8008	119.6701
	F-C-F	4.7997	18.2550	32.3822	47.9172	79.3237	119.1096
	F-F-S	19.1076	37.6454	76.5323	80.7469	147.1610	174.5975
3.0	C-C-C	115.7888	179.7093	249.0593	298.5896	381.3997	518.2542
	C-C-S	87.6011	145.2643	220.2487	263.6977	357.8894	460.4278
	C-C-F	28.4362	57.6522	94.0757	129.9789	204.8213	269.8135
	S-S-S	60.2759	113.6065	180.7275	215.6733	325.3834	427.9401
	S-C-S	72.8739	128.2040	201.1019	242.4808	337.6979	443.0832
	S-C-F	15.6497	39.7599	68.3939	110.7890	175.8428	213.3232
	S-F-S	20.1804	50.3459	89.6868	126.9935	175.1248	266.7918
	F-F-F	11.2146	23.1188	41.5076	46.2533	83.7330	116.7861
	F-C-F	4.6841	17.8151	31.6018	46.7624	77.4120	116.2391
	F-F-S	18.6471	36.7381	74.6879	78.8009	143.6144	170.3897

BCs and μ are also considered in Tables 7.15–7.20. On the other hand, Tables 7.21–7.23 involves the similar edge conditions as stated in Tables 7.12–7.14. Moreover, Tables 7.21–7.23 are meant for the eigenfrequencies of FG triangular elements mentioned in Fig. 7.2(b)–(d), respectively. It can be noticed here that the non-dimensional frequencies are showing descending behavior with ascending values of E_r. This proposition is true because of the fact that the Young's modulus (E_c) is inversely proportional to λ, as observed in the expression concerned.

7.4.3 Effect of the Ratio of Mass Densities (ρ_r)

In this section, Tables 7.24–7.32 show the effect of ascending values of the ratio of mass densities ($\rho_r = 1.0, 1.5, 2.0, 2.5, 3.0$) on free vibration frequencies (λ) of FG triangular plates. The triangular plate in Fig. 7.2(a) is assumed in Tables 7.24–7.29 along with the same set of values for μ and boundary supports, as supposed in Tables 7.6–7.11. Similarly, the non-dimensional frequencies of the plates stated in Fig. 7.2(b)–(d) are reported in Tables 7.30–7.32, respectively. The similar BCs are assumed in Tables 7.30–7.32, as mentioned in Tables 7.12–7.14. The common numerical factors in these computations are $k = 1.0$ and $E_r = 1.0$. It is also interesting to note here that frequencies are increasing with an increase in ρ_r. It may be true because of the fact that λ is directly proportional to ρ_c, as observed in its usual expression. One can also interpret the eigenfrequencies of the FG plate with other sets of BCs after looking into the present results.

Rather than the effects of k, E_r, and ρ_r in Tables 7.6–7.32, there is also another parameter, viz. μ, which has a major impact on the non-dimensional frequencies of FG triangular plates. As such, it is important to observe the descending pattern of frequencies with the ascending values of μ for triangular plate of Fig. 7.2(a).

The results for triangular plates are used here to demonstrate the three-dimensional (or 3D) mode shapes corresponding to the non-dimensional frequencies. As such, Figs. 7.3–7.6 represent the 3D mode shapes of triangular plates corresponding to the first six smallest eigenvalues. Three-dimensional mode shapes of the isotropic equilateral triangular plate ($k = 0$ in the case of FG plate) are considered here in Figs. 7.3 and 7.4 with the edge supports C-C-C and C-F-F, respectively. On the other hand, Figs. 7.5 and 7.6 are meant for the first six mode shapes of equilateral FG plate with $k = 1$, $E_r = 2.0$, and $\rho_r = 1.0$ subjected to C-C-C and C-F-F edge conditions, respectively. Similarly, one may easily plot the mode shapes for other configurations of the FG triangular plates using different sets of BCs.

Table 7.24 Effect of ρ_r on the first six non-dimensional frequencies of C-C-C FG right-angled triangle (Fig. 7.2(a)) for $\theta = 0$, $k = 1.0$, and $E_r = 1.0$

μ	ρ_r	λ_1	λ_2	λ_3	λ_4	λ_5	λ_6
1.0	1.0	93.8094	159.5177	196.4910	248.4158	299.3326	354.9414
	2.0	108.3217	184.1952	226.8882	286.8458	345.6395	409.8511
	2.5	112.1237	190.6601	234.8516	296.9136	357.7708	424.2362
	3.0	114.8926	195.3685	240.6513	304.2460	366.6060	434.7127
1.5	1.0	65.5234	106.7565	141.6423	169.7267	214.1904	255.4156
	2.0	75.6600	123.2717	163.5544	195.9835	247.3257	294.9286
	2.5	78.3155	127.5984	169.2949	202.8622	256.0065	305.2801
	3.0	80.2495	130.7494	173.4756	207.8719	262.3286	312.8190
2.0	1.0	53.4702	83.7901	119.7814	131.0599	172.6714	189.7042
	2.0	61.7421	96.7525	138.3116	151.3350	199.3838	219.0516
	2.5	63.9091	100.1484	143.1661	156.6466	206.3818	226.7399
	3.0	65.4873	102.6215	146.7016	160.5150	211.4784	232.3393
2.5	1.0	46.9768	71.6368	101.9272	110.1294	153.2250	165.7136
	2.0	54.2441	82.7191	117.6954	127.1665	176.9290	191.3496
	2.5	56.1480	85.6224	121.8264	131.6298	183.1389	198.0656
	3.0	57.5346	87.7368	124.8349	134.8804	187.6615	202.9569
3.0	1.0	42.9629	63.0959	88.4648	102.1192	143.8032	156.2600
	2.0	49.6093	72.8569	102.1503	117.9171	166.0497	180.4336
	2.5	51.3505	75.4140	105.7356	122.0558	171.8777	186.7665
	3.0	52.6186	77.2764	108.3468	125.0700	176.1223	191.3787

Table 7.25 Effect of ρ_r on the first six non-dimensional frequencies of C-C-S FG right-angled triangle (Fig. 7.2(a)) for $\theta = 0$, $k = 1.0$, and $E_r = 1.0$

μ	ρ_r	λ_1	λ_2	λ_3	λ_4	λ_5	λ_6
1.0	1.0	73.4250	133.0584	168.5058	222.6888	276.4235	324.2267
	2.0	84.7839	153.6426	194.5738	257.1389	319.1864	374.3847
	2.5	87.7597	159.0352	201.4030	266.1641	330.3893	387.5250
	3.0	89.9269	162.9626	206.3767	272.7370	338.5483	397.0949
1.5	1.0	51.0324	88.9488	121.1213	150.4670	192.3149	235.8292
	2.0	58.9272	102.7093	139.8588	173.7443	222.0661	272.3122
	2.5	60.9954	106.3142	144.7676	179.8424	229.8603	281.8699
	3.0	62.5017	108.9396	148.3427	184.2836	235.5367	288.8307
2.0	1.0	41.3279	69.3498	101.9907	111.0474	158.8008	189.4286
	2.0	47.7213	80.0782	117.7688	128.2265	183.3673	218.7333

Continued

Table 7.25 Effect of ρ_r on the first six non-dimensional frequencies of C-C-S FG right-angled triangle (Fig. 7.2(a)) for $\theta = 0$, $k = 1.0$, and $E_r = 1.0$—cont'd

μ	ρ_r	λ_1	λ_2	λ_3	λ_4	λ_5	λ_6
	2.5	49.3963	82.8888	121.9022	132.7271	189.8032	226.4105
	3.0	50.6161	84.9358	124.9126	136.0048	194.4904	232.0017
2.5	1.0	36.0380	58.1675	87.5269	94.2710	143.4914	172.7007
	2.0	41.6131	67.1661	101.0674	108.8548	165.6896	199.4176
	2.5	43.0737	69.5235	104.6147	112.6754	171.5051	206.4168
	3.0	44.1374	71.2404	107.1982	115.4580	175.7404	211.5143
3.0	1.0	32.7042	50.5756	77.9265	87.9661	135.7907	165.2824
	2.0	37.7635	58.3996	89.9817	101.5745	156.7976	190.8516
	2.5	39.0890	60.4494	93.1399	105.1396	162.3009	197.5502
	3.0	40.0543	61.9422	95.4400	107.7361	166.3090	202.4287

Table 7.26 Effect of ρ_r on the first six non-dimensional frequencies of S-S-S FG right-angled triangle (Fig. 7.2(a)) for $\theta = 0$, $k = 1.0$, and $E_r = 1.0$

μ	ρ_r	λ_1	λ_2	λ_3	λ_4	λ_5	λ_6
1.0	1.0	49.3772	102.3974	132.8920	181.7691	239.4133	286.9646
	2.0	57.0159	118.2384	153.4505	209.8888	276.4506	331.3581
	2.5	59.0171	122.3883	158.8364	217.2556	286.1536	342.9883
	3.0	60.4745	125.4107	162.7588	222.6207	293.2202	351.4584
1.5	1.0	34.3846	67.1145	96.0449	129.7321	174.5341	203.5525
	2.0	39.7039	77.4972	110.9031	149.8018	201.5346	235.0421
	2.5	41.0975	80.2172	114.7956	155.0596	208.6081	243.2917
	3.0	42.1124	82.1982	117.6305	158.8888	213.7597	249.2998
2.0	1.0	27.8391	51.4624	82.3722	107.7080	143.4821	172.1820
	2.0	32.1458	59.4237	95.1152	124.3704	165.6788	198.8186
	2.5	33.2741	61.5093	98.4536	128.7356	171.4939	205.7968
	3.0	34.0958	63.0283	100.8849	131.9148	175.7290	210.8790
2.5	1.0	24.1904	43.9850	74.6922	89.6844	122.8670	149.5437
	2.0	27.9326	50.7895	86.2471	103.5586	141.8746	172.6782
	2.5	28.9130	52.5721	89.2743	107.1934	146.8541	178.7389
	3.0	29.6270	53.8704	91.4789	109.8405	150.4807	183.1528
3.0	1.0	21.9385	39.6477	69.2333	73.5403	110.9179	132.3405
	2.0	25.3324	45.7813	79.9438	84.9170	128.0769	152.8137
	2.5	26.2216	47.3881	82.7497	87.8974	132.5722	158.1772
	3.0	26.8691	48.5584	84.7932	90.0681	135.8461	162.0834

Table 7.27 Effect of ρ_r on the first six non-dimensional frequencies of C-C-F FG right-angled triangle (Fig. 7.2(a)) for $\theta = 0$, $k = 1.0$, and $E_r = 1.0$

μ	ρ_r	λ_1	λ_2	λ_3	λ_4	λ_5	λ_6
1.0	1.0	29.0988	63.8764	90.4624	118.4707	171.1197	214.0724
	2.0	33.6003	73.7581	104.4570	136.7982	197.5921	247.1895
	2.5	34.7797	76.3469	108.1232	141.5995	204.5272	255.8654
	3.0	35.6386	78.2323	110.7934	145.0964	209.5780	262.1841
1.5	1.0	19.7848	41.7050	64.9056	77.9933	118.6126	142.0786
	2.0	22.8455	48.1568	74.9465	90.0589	136.9620	164.0582
	2.5	23.6474	49.8470	77.5770	93.2198	141.7691	169.8164
	3.0	24.2313	51.0780	79.4928	95.5218	145.2701	174.0100
2.0	1.0	15.4506	30.6559	53.8549	59.1377	90.3218	99.9366
	2.0	17.8408	35.3983	62.1863	68.2863	104.2946	115.3969
	2.5	18.4670	36.6408	64.3690	70.6830	107.9552	119.4471
	3.0	18.9230	37.5456	65.9586	72.4286	110.6212	122.3969
2.5	1.0	12.9612	24.6949	43.5926	50.2031	74.7484	85.6615
	2.0	14.9663	28.5152	50.3364	57.9696	86.3120	98.9134
	2.5	15.4916	29.5161	52.1031	60.0042	89.3414	102.3851
	3.0	15.8741	30.2450	53.3898	61.4860	91.5477	104.9135
3.0	1.0	11.3563	21.0540	35.0606	45.9334	68.1932	78.6279
	2.0	13.1131	24.3111	40.4845	53.0393	78.7428	90.7917
	2.5	13.5734	25.1643	41.9054	54.9009	81.5065	93.9783
	3.0	13.9086	25.7858	42.9403	56.2567	83.5193	96.2991

Table 7.28 Effect of ρ_r on the first six non-dimensional frequencies of C-S-F FG right-angled triangle (Fig. 7.2(a)) for $\theta = 0$, $k = 1.0$, and $E_r = 1.0$

μ	ρ_r	λ_1	λ_2	λ_3	λ_4	λ_5	λ_6
1.0	1.0	17.9703	48.1308	75.7060	100.7389	155.2088	200.8066
	2.0	20.7503	55.5767	87.4178	116.3233	179.2197	231.8715
	2.5	21.4786	57.5273	90.4860	120.4060	185.5100	240.0098
	3.0	22.0090	58.9480	92.7206	123.3795	190.0912	245.9369
1.5	1.0	13.7573	34.3827	54.4613	69.2721	110.9556	142.1440
	2.0	15.8856	39.7017	62.8865	79.9885	128.1205	164.1338
	2.5	16.4432	41.0952	65.0937	82.7960	132.6174	169.8946
	3.0	16.8492	42.1100	66.7012	84.8407	135.8924	174.0902
2.0	1.0	11.5280	26.2811	45.9373	53.7887	92.0098	118.5867
	2.0	13.3114	30.3468	53.0438	62.1099	106.2437	136.9321

Continued

Table 7.28 Effect of ρ_r on the first six non-dimensional frequencies of C-S-F FG right-angled triangle (Fig. 7.2(a)) for $\theta = 0$, $k = 1.0$, and $E_r = 1.0$—cont'd

μ	ρ_r	λ_1	λ_2	λ_3	λ_4	λ_5	λ_6
	2.5	13.7786	31.4119	54.9056	64.2898	109.9727	141.7382
	3.0	14.1189	32.1877	56.2615	65.8775	112.6885	145.2384
2.5	1.0	10.1513	21.2417	41.0766	46.8536	78.6180	96.0617
	2.0	11.7217	24.5278	47.4312	54.1018	90.7802	110.9225
	2.5	12.1331	25.3887	49.0959	56.0007	93.9665	114.8157
	3.0	12.4327	26.0157	50.3084	57.3837	96.2870	117.6510
3.0	1.0	9.2021	18.1554	37.5761	42.9889	68.0010	75.4984
	2.0	10.6257	20.9641	43.3891	49.6393	78.5208	87.1781
	2.5	10.9986	21.6999	44.9120	51.3816	81.2768	90.2379
	3.0	11.2702	22.2358	46.0211	52.6505	83.2839	92.4663

Table 7.29 Effect of ρ_r on the first six non-dimensional frequencies of C-F-F FG right-angled triangle (Fig. 7.2(a)) for $\theta = 0$, $k = 1.0$, and $E_r = 1.0$

μ	ρ_r	λ_1	λ_2	λ_3	λ_4	λ_5	λ_6
1.0	1.0	6.1807	23.5033	32.8433	59.7726	82.9699	116.0983
	2.0	7.1369	27.1392	37.9241	69.0195	95.8054	134.0588
	2.5	7.3874	28.0918	39.2552	71.4420	99.1680	138.7640
	3.0	7.5698	28.7855	40.2246	73.2062	101.6170	142.1908
1.5	1.0	5.8011	18.4980	29.8889	42.2561	71.2843	87.6627
	2.0	6.6986	21.3597	34.5127	48.7931	82.3120	101.2242
	2.5	6.9337	22.1094	35.7240	50.5057	85.2010	104.7770
	3.0	7.1049	22.6554	36.6062	51.7529	87.3050	107.3645
2.0	1.0	5.5182	15.4110	28.6530	32.4307	63.4524	80.1162
	2.0	6.3719	17.7951	33.0856	37.4477	73.2685	92.5102
	2.5	6.5955	18.4197	34.2468	38.7621	75.8401	95.7571
	3.0	6.7584	18.8745	35.0926	39.7193	77.7130	98.1218
2.5	1.0	5.3074	13.3286	27.5323	27.9591	57.7683	73.6492
	2.0	6.1285	15.3906	31.7915	32.2843	66.7051	85.0428
	2.5	6.3436	15.9307	32.9074	33.4175	69.0464	88.0276
	3.0	6.5002	16.3242	33.7200	34.2427	70.7515	90.2015
3.0	1.0	5.1428	11.8874	24.8818	27.3406	53.3694	57.8147
	2.0	5.9384	13.7264	28.7311	31.5702	61.6257	66.7587
	2.5	6.1468	14.2082	29.7395	32.6783	63.7886	69.1018
	3.0	6.2986	14.5590	30.4739	33.4852	65.3639	70.8083

Table 7.30 Effect of ρ_r on the first six non-dimensional frequencies of equilateral FG triangular plates (Fig. 7.2(b)) with $\theta = 1/\sqrt{3}$ and $\mu = \sqrt{3}/2$

ρ_r	BCs	λ_1	λ_2	λ_3	λ_4	λ_5	λ_6
1.0	C-C-C	99.0301	189.3558	189.3558	302.5013	320.2886	320.2886
	C-C-S	81.6095	165.6640	165.9714	274.0097	295.8453	299.9950
	C-C-F	40.0393	95.9871	102.0821	176.7503	204.5401	206.2233
	S-S-S	52.6389	124.5893	124.5893	219.4339	243.2077	243.2077
	S-C-S	66.1950	143.8823	144.4059	253.5495	269.0919	273.7237
	S-C-F	26.5733	75.6211	84.9659	152.7500	182.8633	185.6936
	S-F-S	16.0939	58.1982	68.7730	124.7878	156.2157	158.3566
	F-F-F	35.2728	36.5976	36.5976	111.9604	111.9604	118.7939
	F-C-F	8.9250	35.3952	38.5490	95.6379	99.2576	115.4143
	F-F-S	22.6792	26.8547	72.2202	75.6233	92.9068	190.9280
2.0	C-C-C	114.3501	218.6493	218.6493	349.2984	369.8374	369.8374
	C-C-S	94.2345	191.2923	191.6472	316.3991	341.6127	346.4044
	C-C-F	46.2334	110.8364	117.8742	204.0936	236.1826	238.1262
	S-S-S	60.7821	143.8633	143.8633	253.3805	280.8320	280.8320
	S-C-S	76.4354	166.1409	166.7455	292.7738	310.7206	316.0689
	S-C-F	30.6842	87.3197	98.1101	176.3806	211.1523	214.4205
	S-F-S	18.5837	67.2015	79.4122	144.0925	180.3824	182.8545
	F-F-F	40.7295	42.2593	42.2593	129.2807	129.2807	137.1714
	F-C-F	10.3057	40.8708	44.5126	110.4331	114.6128	133.2690
	F-F-S	26.1876	31.0091	83.3927	87.3223	107.2796	220.4647
2.5	C-C-C	118.3636	226.3235	226.3235	361.5582	382.8181	382.8181
	C-C-S	97.5420	198.0064	198.3737	327.5042	353.6027	358.5626
	C-C-F	47.8561	114.7265	122.0114	211.2570	244.4722	246.4840
	S-S-S	62.9155	148.9127	148.9127	262.2737	290.6888	290.6888
	S-C-S	79.1182	171.9722	172.5980	303.0497	321.6263	327.1624
	S-C-F	31.7612	90.3845	101.5536	182.5712	218.5634	221.9463
	S-F-S	19.2359	69.5602	82.1995	149.1499	186.7135	189.2724
	F-F-F	42.1591	43.7425	43.7425	133.8182	133.8182	141.9859
	F-C-F	10.6674	42.3053	46.0749	114.3092	118.6355	137.9465
	F-F-S	27.1068	32.0974	86.3197	90.3871	111.0449	228.2026
3.0	C-C-C	121.2866	231.9126	231.9126	370.4869	392.2718	392.2718
	C-C-S	99.9508	202.8961	203.2726	335.5919	362.3350	367.4173
	C-C-F	49.0380	117.5597	125.0245	216.4740	250.5094	252.5710
	S-S-S	64.4692	152.5901	152.5901	268.7506	297.8673	297.8673
	S-C-S	81.0720	176.2191	176.8603	310.5335	329.5689	335.2417
	S-C-F	32.5455	92.6166	104.0615	187.0798	223.9609	227.4273
	S-F-S	19.7109	71.2780	84.2294	152.8332	191.3244	193.9465
	F-F-F	43.2002	44.8227	44.8227	137.1229	137.1229	145.4923
	F-C-F	10.9308	43.3500	47.2127	117.1320	121.5652	141.3531
	F-F-S	27.7762	32.8901	88.4514	92.6193	113.7872	233.8381

Table 7.31 Effect of ρ_r on the first six non-dimensional frequencies of FG right-angled triangular plates (Fig. 7.2(c)) with angles 30°, 60°, and 90°

ρ_r	BCs	λ_1	λ_2	λ_3	λ_4	λ_5	λ_6
1.0	C-C-C	176.6773	281.6946	389.4979	449.9239	574.4351	657.6405
	C-C-S	137.1100	234.4229	332.2733	388.2433	520.0610	626.8953
	C-C-F	52.2997	107.0327	177.4260	203.2358	312.6933	349.1134
	S-S-S	92.4337	174.8875	265.3512	354.5446	476.1434	556.7795
	S-C-S	119.5542	213.0784	305.1774	376.5482	523.1429	614.7466
	S-C-F	37.7140	90.4292	149.3749	181.3869	302.9555	388.3029
	S-F-S	22.9903	66.3451	130.0744	144.0188	287.5747	356.4993
	F-F-F	25.5779	54.5909	81.0811	101.2828	143.6803	231.9947
	F-C-F	16.9768	50.6991	87.6860	110.4150	202.0306	249.3552
	F-F-S	32.0453	69.2456	89.0420	170.3531	224.4979	241.4710
2.0	C-C-C	204.0093	325.2729	449.7534	519.5274	663.3006	759.3779
	C-C-S	158.3210	270.6882	383.6761	448.3048	600.5147	723.8764
	C-C-F	60.3905	123.5907	204.8739	234.6765	361.0671	403.1214
	S-S-S	106.7333	201.9427	306.4012	409.3929	549.8031	642.9136
	S-C-S	138.0493	246.0417	352.3886	434.8004	604.0734	709.8482
	S-C-F	43.5484	104.4187	172.4833	209.4475	349.8228	448.3735
	S-F-S	26.5470	76.6087	150.1970	166.2986	332.0626	411.6500
	F-F-F	29.5348	63.0362	93.6243	116.9514	165.9077	267.8844
	F-C-F	19.6031	58.5423	101.2511	127.4963	233.2849	287.9306
	F-F-S	37.0028	79.9580	102.8168	196.7069	259.2278	278.8267
2.5	C-C-C	211.1697	336.6894	465.5390	537.7619	686.5813	786.0308
	C-C-S	163.8778	280.1889	397.1425	464.0395	621.5917	749.2832
	C-C-F	62.5101	127.9286	212.0647	242.9132	373.7399	417.2703
	S-S-S	110.4795	209.0306	317.1554	423.7619	569.1003	665.4788
	S-C-S	142.8947	254.6774	364.7568	450.0612	625.2753	734.7627
	S-C-F	45.0769	108.0836	178.5371	216.7988	362.1010	464.1107
	S-F-S	27.4787	79.2975	155.4686	172.1354	343.7175	426.0982
	F-F-F	30.5715	65.2486	96.9104	121.0561	171.7308	277.2867
	F-C-F	20.2912	60.5971	104.8048	131.9712	241.4728	298.0365
	F-F-S	38.3015	82.7644	106.4255	203.6109	268.3263	288.6131
3.0	C-C-C	216.3846	345.0040	477.0355	551.0420	703.5365	805.4419
	C-C-S	167.9248	287.1082	406.9500	475.4990	636.9420	767.7868
	C-C-F	64.0538	131.0878	217.3016	248.9120	382.9695	427.5749
	S-S-S	113.2078	214.1926	324.9876	434.2267	583.1542	681.9129
	S-C-S	146.4234	260.9667	373.7645	461.1755	640.7166	752.9077
	S-C-F	46.1901	110.7527	182.9461	222.1526	371.0432	475.5719
	S-F-S	28.1573	81.2558	159.3080	176.3863	352.2056	436.6207
	F-F-F	31.3264	66.8600	99.3036	124.0456	175.9717	284.1343
	F-C-F	20.7923	62.0935	107.3930	135.2303	247.4360	305.3965
	F-F-S	39.2474	84.8082	109.0537	208.6391	274.9526	295.7404

Table 7.32 Effect of ρ_r on the first six non-dimensional frequencies of isosceles FG triangular plates (Fig. 7.2(d)) with angles 30°, 30°, and 120°

ρ_r	BCs	λ_1	λ_2	λ_3	λ_4	λ_5	λ_6
1.0	C-C-C	141.8117	220.0980	305.0341	365.6960	467.1174	634.7292
	C-C-S	107.2890	177.9117	269.7484	322.9624	438.3232	563.9066
	C-C-F	34.8270	70.6093	115.2188	159.1910	250.8538	330.4526
	S-S-S	73.8226	139.1389	221.3450	264.1448	398.5116	524.1175
	S-C-S	89.2520	157.0172	246.2985	296.9771	413.5938	542.6639
	S-C-F	19.1669	48.6957	83.7651	135.6883	215.3625	261.2665
	S-F-S	24.7159	61.6608	109.8434	155.5346	214.4833	326.7519
	F-F-F	13.7350	28.3146	50.8362	56.6484	102.5516	143.0332
	F-C-F	5.7368	21.8189	38.7042	57.2720	94.8100	142.3632
	F-F-S	22.8380	44.9948	91.4736	96.5110	175.8910	208.6840
2.0	C-C-C	163.7501	254.1473	352.2230	422.2694	539.3807	732.9222
	C-C-S	123.8866	205.4347	311.4786	372.9249	506.1320	651.1432
	C-C-F	40.2148	81.5326	133.0432	183.8180	289.6611	381.5738
	S-S-S	85.2430	160.6638	255.5872	305.0081	460.1616	605.1987
	S-C-S	103.0593	181.3079	284.4010	342.9196	477.5769	626.6143
	S-C-F	22.1321	56.2290	96.7236	156.6794	248.6792	301.6845
	S-F-S	28.5394	71.1998	126.8363	179.5959	247.6639	377.3006
	F-F-F	15.8599	32.6949	58.7006	65.4120	118.4164	165.1605
	F-C-F	6.6243	25.1943	44.6917	66.1320	109.4771	164.3869
	F-F-S	26.3710	51.9555	105.6247	111.4413	203.1014	240.9675
2.5	C-C-C	169.4975	263.0674	364.5854	437.0904	558.3120	758.6465
	C-C-S	128.2349	212.6451	322.4110	386.0140	523.8964	673.9973
	C-C-F	41.6263	84.3942	137.7128	190.2697	299.8277	394.9665
	S-S-S	88.2349	166.3028	264.5579	315.7134	476.3125	626.4402
	S-C-S	106.6765	187.6715	294.3830	354.9555	494.3391	648.6074
	S-C-F	22.9089	58.2025	100.1184	162.1785	257.4075	312.2732
	S-F-S	29.5411	73.6988	131.2880	185.8995	256.3565	390.5432
	F-F-F	16.4165	33.8424	60.7609	67.7078	122.5726	170.9573
	F-C-F	6.8568	26.0786	46.2603	68.4532	113.3196	170.1566
	F-F-S	27.2966	53.7791	109.3319	115.3527	210.2299	249.4250
3.0	C-C-C	173.6832	269.5639	373.5889	447.8844	572.0996	777.3814
	C-C-S	131.4016	217.8964	330.3730	395.5466	536.8340	690.6417
	C-C-F	42.6542	86.4784	141.1136	194.9684	307.2319	404.7202
	S-S-S	90.4139	170.4097	271.0912	323.5099	488.0751	641.9102
	S-C-S	109.3109	192.3061	301.6528	363.7212	506.5468	664.6248
	S-C-F	23.4746	59.6398	102.5909	166.1836	263.7642	319.9848
	S-F-S	30.2706	75.5188	134.5302	190.4903	262.6873	400.1877
	F-F-F	16.8219	34.6782	62.2614	69.3799	125.5995	175.1791
	F-C-F	7.0261	26.7226	47.4027	70.1436	116.1180	174.3586
	F-F-S	27.9707	55.1072	112.0319	118.2013	215.4216	255.5846

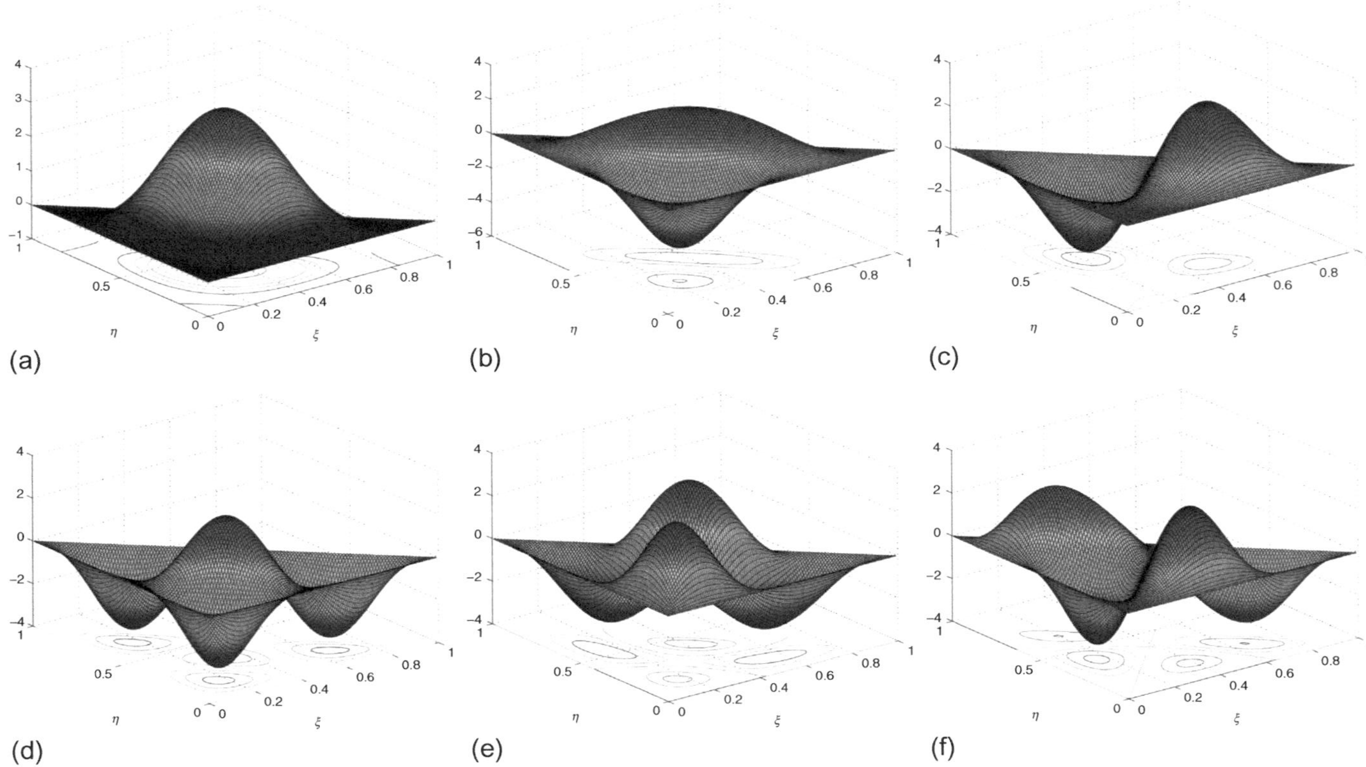

Figure 7.3 The first six 3D mode shapes of C-C-C equilateral triangular plate with $k = 0$.

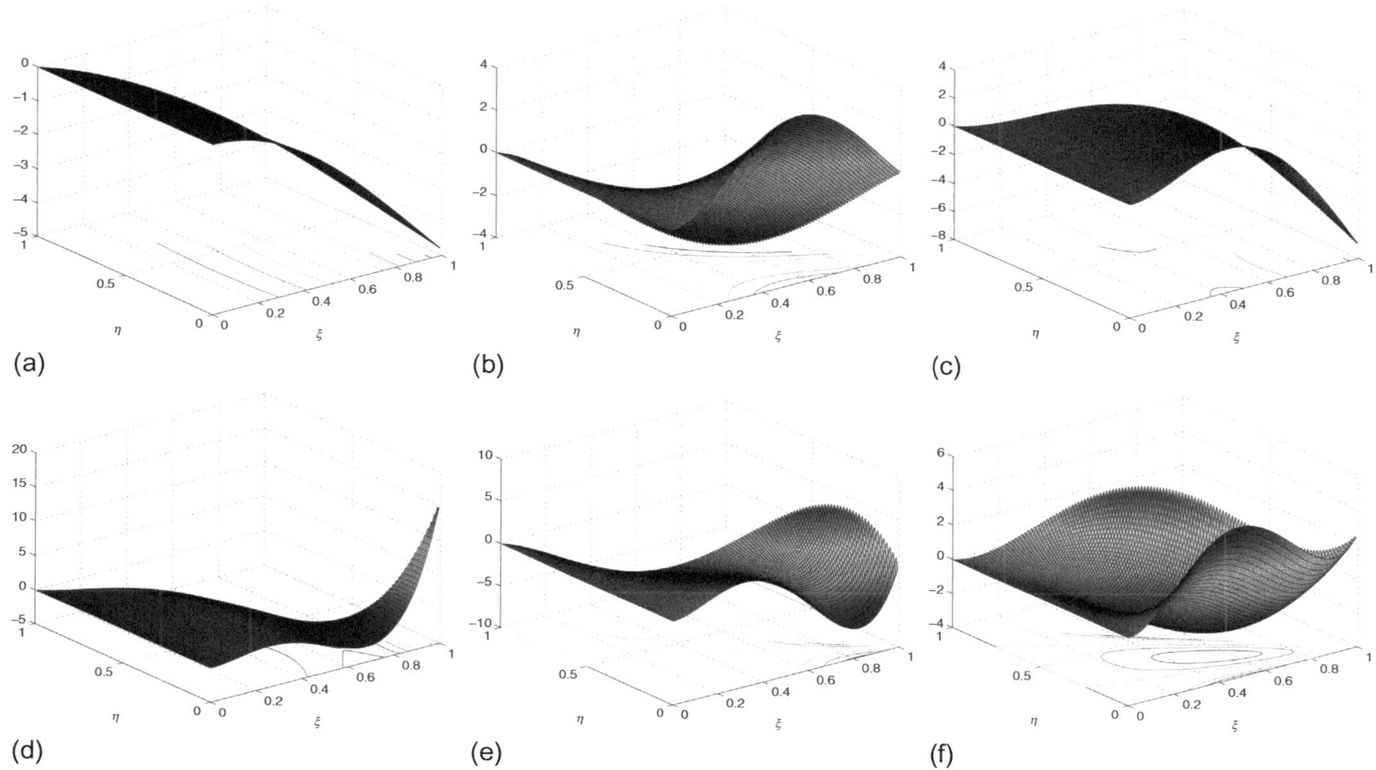

Figure 7.4 The first six 3D mode shapes of C-F-F equilateral triangular plate with $k = 0$.

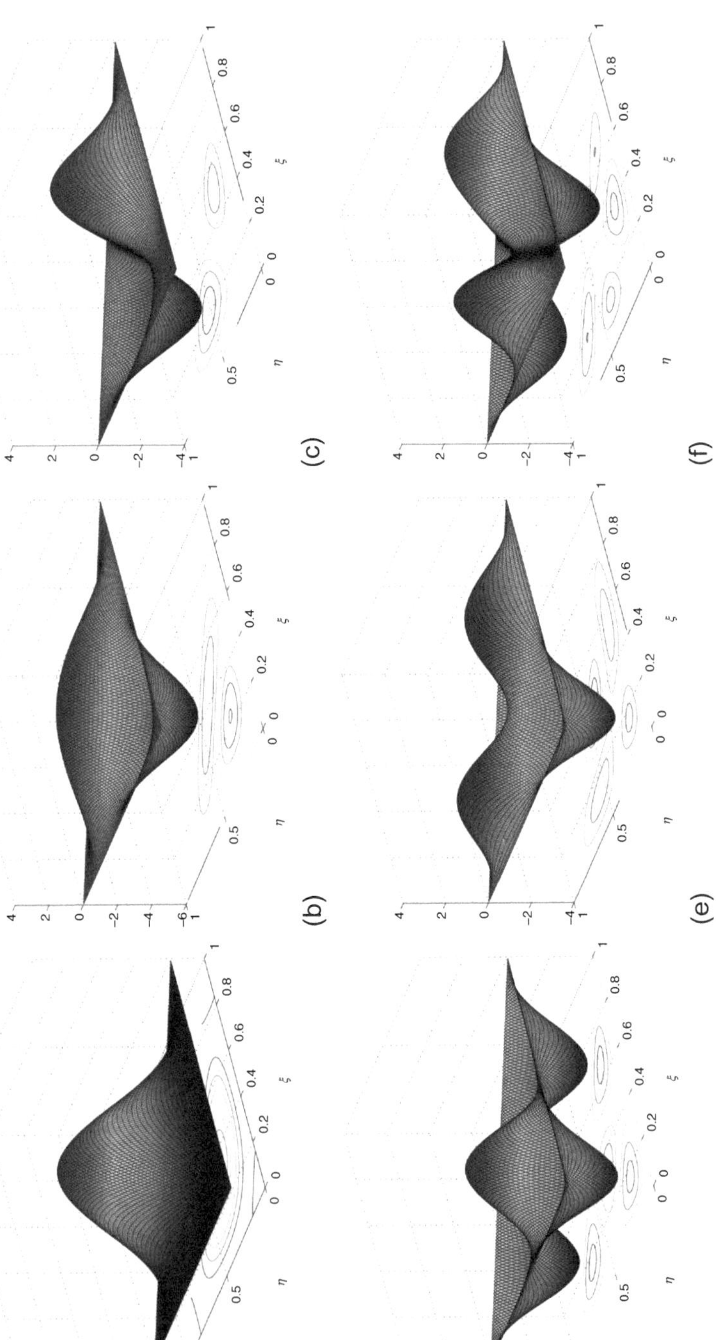

Figure 7.5 The first six 3D mode shapes of C-C-C FG equilateral triangular plate with $k = 1$, $E_r = 2.0$, and $\rho_r = 1.0$.

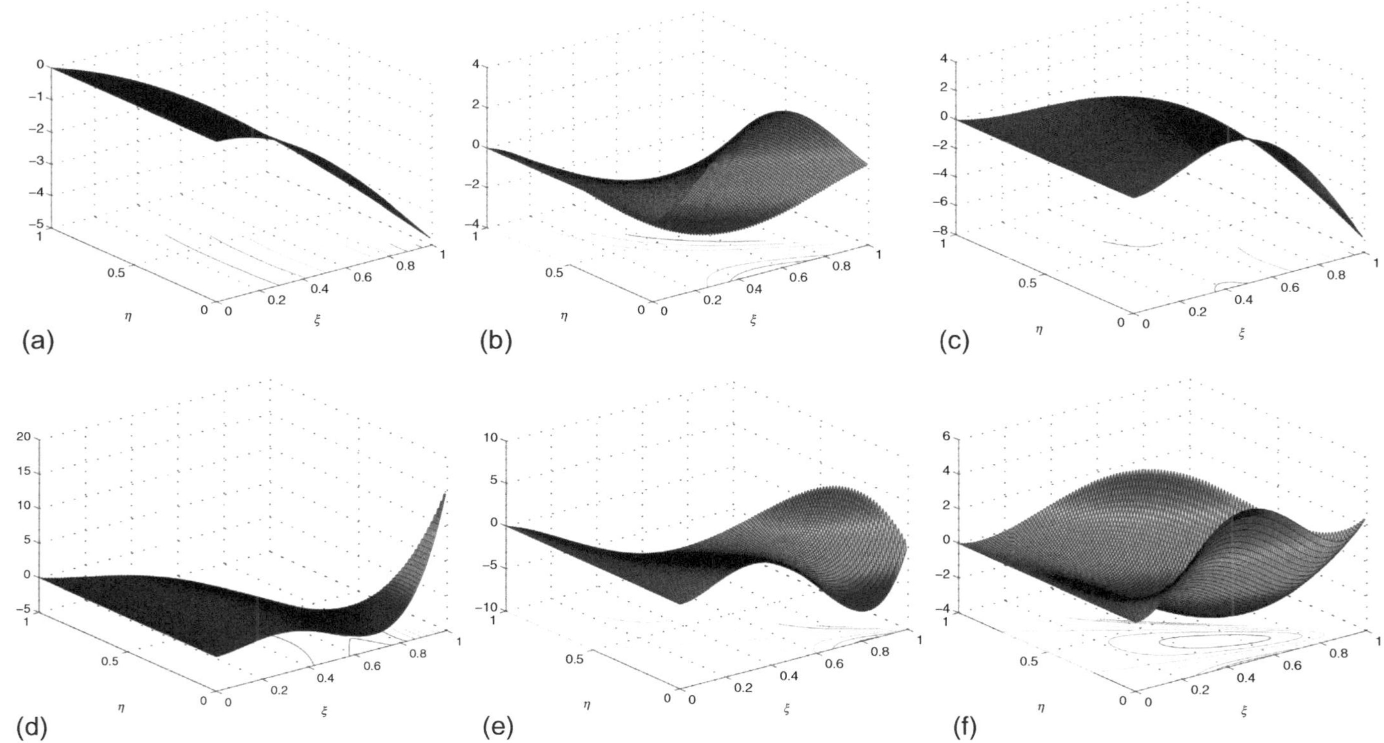

Figure 7.6 The first six 3D mode shapes of C-F-F FG equilateral triangular plate with $k = 1, E_r = 2.0$, and $\rho_r = 1.0$.

7.5 CONCLUDING REMARKS

This investigation deals especially with free vibration of FG triangular plates subjected to various combinations of BCs. Free vibration eigenfrequencies along with 3D mode shapes have been incorporated with reference to different volume fractions of the FG plate. It may be worth mentioning that the Rayleigh-Ritz method is one of the most computationally efficient techniques, which may handle any combinations of BCs very easily. We hope that the present results may serve as a benchmark for other researchers in future investigations. From the above numerical results and discussions, one may summarize the following facts for the free vibration of FG triangular plates.

- The present method assumes the simple algebraic polynomials as the admissible functions for computing the frequencies and mode shapes of a triangular plate with any set of BCs. The geometric configuration and BCs can be controlled by five parameters, viz. p, q, r, μ, and θ.
- The method can easily be extended to other geometries by a suitable selection of the admissible function. Also, suitable modifications can be made to include variable thickness and other complicating effects.
- An increase in the number of polynomials (n) plays an important role in the convergence of non-dimensional frequencies. Moreover, different volume fractions developed from the parameters, viz. k, E_r, and ρ_r, also have major roles in the variation of frequencies. Each of the eigenfrequencies follow a descending pattern due to an increase in k and E_r, whereas we get just a reverse pattern (ascending behavior) with an increase in ρ_r. Rather than these parameters, the factor μ also plays a crucial role in evaluating vibration characteristics of the FG triangular plates (Fig. 7.2(a)). It may be viewed (in this case) that frequencies are decreasing with an increase in μ.
- The FG plate behaves as isotropic either for $k = 0$ or for the combined values of $E_r = 1.0$ and $\rho_r = 1.0$.
- In addition to the shear deformation effects (neglected in classical plate theory), the analysis may also be extended for other deformation plate theories.

CHAPTER 8

Complicating Effects

As mentioned earlier, the concept of functionally graded (FG) materials is derived to withstand a huge temperature fluctuation. In real-life applications, the structural members may be in various complicating environments, viz. thermal effects, elastic foundations, piezoelectric effects, etc., which play a major role in judging vibration characteristics of isotropic as well as FG plates. In this regard, the present literature helps us to find a variety of investigations performed in these domains.

Let us first discuss the studies on the effect of elastic foundation on vibration characteristics of plates. As such, the elastic bending, buckling, and vibration of Lévy plates resting on various two-parameter elastic foundations is investigated by Lam et al. (2000). The dynamic response of initially stressed FG rectangular thin plates is obtained by Yang and Shen (2001), subjected to partially distributed impulsive lateral loads and without or resting on an elastic foundation. Hosseini-Hashemi and Arsanjani (2005) have presented the exact closed form characteristic equations for the Mindlin plates with two opposite edges simply supported. The exact solutions for free vibration analysis of rectangular Mindlin plates have been obtained by Akhavan et al. (2009) under in-plane loads resting on a Pasternak elastic foundation. Free vibration analysis of vertical rectangular Mindlin plates resting on a Pasternak elastic foundation is given in Hosseini-Hashemi et al. (2010) for different combinations of boundary conditions (BCs). Baferani et al. (2011) have presented vibration analysis of FG rectangular plates resting on two parameter elastic foundations based on third-order shear deformation theory. Flexural vibration of Lévy-type rectangular plates has been studied by Hosseini-Hashemi et al. (2011) within the framework of third-order shear deformation theory. A refined shear deformation theory is considered by Thai and Choi (2012) for free vibration of FG plates on elastic foundation. Fallah et al. (2013) have studied free vibration analysis of moderately thick FG rectangular plates on elastic foundations.

Vibration of Functionally Graded Beams and Plates
http://dx.doi.org/10.1016/B978-0-12-804228-1.00008-0

Next, we include different works done on mechanics of structural members in thermal environments. Based on first-order shear deformation theory (FSDT), Reddy and Chin (1998) have analyzed the dynamic thermoelastic response of FG cylinders and plates using thermomechanical coupling. Nonlinear bending analysis is presented by Shen (2002) for a simply supported FG plate subjected to a transverse uniform or sinusoidal loads in thermal environments. Yang and Shen (2002) have presented free and forced vibration analyses for initially stressed FG plates in a thermal environment. One may also see Kim (2005), Kitipornchai et al. (2006), Li et al. (2009a), Malekzadeh and Beni (2010), and Shi and Dong (2012) and additionally the literatures therein to associate the study with the effect of thermal or temperature environment on vibration analysis of FG plates. An analytical model is developed by Bouchafa et al. (2010) for prediction of thermal residual stresses, arising from the fabrication of exponential FG material systems. Considering the complicating effects, Chakraverty and Pradhan (2014a,b) have investigated free vibration of thin FG rectangular plates in very recent literature.

This chapter reflects the effect of different complicating effects, viz. Winkler (or Pasternak) elastic foundations and thermal environments on vibration characteristics of FG rectangular plates. The presence of complicating environments mainly manipulates the expression of strain energy added with a specific term. Against this, the basic approach of the Rayleigh-Ritz method (as mentioned in Chapter 3) is also considered in these problems. It can further be seen that the factors associated with these complicating environments play major roles in finding free vibration of rectangular FG plates. Finally, a short review on vibration of FG beams and plates integrated with piezoelectric patches is performed, along with consideration of their significance in various engineering and industrial sectors. In addition, the equilibrium equations for piezoelectric effects are also given.

8.1 WINKLER AND PASTERNAK FOUNDATIONS

In this section, free vibration of thin FG Lévy plates resting on Winkler and Pasternak elastic foundations has been investigated. A simple power-law form mentioned in Eq. (2.1) is also assumed here to judge the variation of material properties of the FG constituents along the thickness direction. The prime objective is to find the effect of different physical and geometrical parameters along with elastic foundations on free vibration characteristics of FG Lévy plates. Here also, the results of eigenfrequencies and their corresponding 3D mode shapes are incorporated after performing the test of convergence and validation of present results with the available literature in special cases.

8.1.1 Numerical Modeling

The expressions of strain and kinetic energies of FG Lévy plates in the presence of Winkler and Pasternak elastic foundations can be written as

$$\mathbf{U} = \frac{1}{2}\int_{\Omega}\left[D_{11}\left\{\left(\frac{\partial^2 w}{\partial x^2}\right)^2 + \left(\frac{\partial^2 w}{\partial y^2}\right)^2\right\} + 2D_{12}\frac{\partial^2 w}{\partial x^2}\frac{\partial^2 w}{\partial y^2} + 4D_{66}\left(\frac{\partial^2 w}{\partial x \partial y}\right)^2 \right.$$
$$\left. + k_{\mathrm{w}} w^2 + k_{\mathrm{p}}\left\{\left(\frac{\partial w}{\partial x}\right)^2 + \left(\frac{\partial w}{\partial y}\right)^2\right\}\right] \mathrm{d}x\,\mathrm{d}y \tag{8.1}$$

$$\mathbf{T} = \frac{1}{2}\int_{\Omega} I_0 \left(\frac{\partial w}{\partial t}\right)^2 \mathrm{d}x\,\mathrm{d}y \tag{8.2}$$

where k_{w} and k_{p} represent the Winkler and Pasternak foundation moduli respectively; the stiffness coefficients $(D_{ij};\ i, j = 1, 2, 6)$ are

$$(D_{11}, D_{12}, D_{66}) = \int_{-h/2}^{h/2} (Q_{11}, Q_{12}, Q_{66}) z^2 \,\mathrm{d}z$$

and inertial coefficient, $I_0 = \int_{-h/2}^{h/2} \rho(z)\,\mathrm{d}z$.

We assume harmonic type displacement as $w(x, y, t) = W(x, y)\cos\omega t$, where $W(x, y)$ and ω are respective maximum deflection and natural frequency of free vibration. Accordingly, Eqs. (8.1) and (8.2) may be transformed into maximum strain energy ($\mathbf{U}_{\max}$) and kinetic energy ($\mathbf{T}_{\max}$) respectively as follows:

$$\mathbf{U}_{\max} = \frac{1}{2}\int_{\Omega}\left[D_{11}\left\{\left(\frac{\partial^2 W}{\partial x^2}\right)^2 + \left(\frac{\partial^2 W}{\partial y^2}\right)^2\right\} + 2D_{12}\frac{\partial^2 W}{\partial x^2}\frac{\partial^2 W}{\partial y^2} \right.$$
$$\left. + 4D_{66}\left(\frac{\partial^2 W}{\partial x \partial y}\right)^2 + k_{\mathrm{w}} W^2 + k_{\mathrm{p}}\left\{\left(\frac{\partial W}{\partial x}\right)^2 + \left(\frac{\partial W}{\partial y}\right)^2\right\}\right] \mathrm{d}x\,\mathrm{d}y$$

$$\mathbf{T}_{\max} = \frac{\omega^2}{2}\int_{\Omega} I_0 W^2 \,\mathrm{d}x\,\mathrm{d}y$$

Let us now consider the cartesian coordinates x and y as the non-dimensionalized parameters ($\xi = x/a$) and ($\eta = y/b$) respectively in natural coordinate system. Consequently, the expressions for maximum strain and kinetic energies can be written as

$$\mathbf{U}_{\max} = \frac{D_c ab}{2a^4} \int_{\Omega} \left[\overline{D}_{11} \left\{ \left(\frac{\partial^2 W}{\partial \xi^2} \right)^2 + \mu^4 \left(\frac{\partial^2 W}{\partial \eta^2} \right)^2 \right\} + 2\overline{D}_{12}\mu^2 \frac{\partial^2 W}{\partial \xi^2} \frac{\partial^2 W}{\partial \eta^2} \right.$$
$$+ 4\overline{D}_{66}\mu^2 \left(\frac{\partial^2 W}{\partial \xi \partial \eta} \right)^2 + K_w W^2$$
$$\left. + K_p \left\{ \left(\frac{\partial W}{\partial \xi} \right)^2 + \mu^2 \left(\frac{\partial W}{\partial \eta} \right)^2 \right\} \right] \mathrm{d}\xi \, \mathrm{d}\eta \tag{8.3}$$

$$\mathbf{T}_{\max} = \frac{\rho_c h \omega^2 ab}{2} \int_{\Omega} I_0 W^2 \, \mathrm{d}\xi \, \mathrm{d}\eta \tag{8.4}$$

where the flexural rigidity $D_c = \frac{\rho_c h^3}{12(1-\nu^2)}$ and other components in Eqs. (8.3) and (8.4) are $\overline{D}_{11} = \frac{D_{11}}{D_c}$, $\overline{D}_{12} = \frac{D_{12}}{D_c}$, $\overline{D}_{66} = \frac{D_{66}}{D_c}$, $\mu = \frac{a}{b}$, $K_w = \frac{k_w a^4}{D_c}$, and $K_p = \frac{k_p a^2}{D_c}$.

The amplitude $W(\xi, \eta)$ may be expressed as the sum of series of simple algebraic polynomials involving both ξ and η as follows:

$$W(\xi, \eta) = \sum_{i=1}^{n} c_i \varphi_i(\xi, \eta)$$

where c_i are unknown constants to be determined and $\varphi_i(\xi, \eta) = f\psi_i(\xi, \eta)$, $i = 1, 2, \ldots, n$ are the admissible functions. Here, n is the number of polynomials involved in the admissible functions. The function $f = \xi^p \eta^q (1-\xi)^r (1-\eta)^s$, with the exponents p, q, r, and s, controls various BCs. The parameter $p = 0$, 1, or 2 according as the side $\xi = 0$ is free (F), simply supported (S), or clamped (C). Similar interpretations can be given to the parameters q, r, and s corresponding to the sides $\eta = 0$, $\xi = 1$, and $\eta = 1$, respectively. The components of ψ_i will be generated from Pascal's triangle as given in Table 3.2.

Assuming constant Poisson's ratio (ν), the Rayleigh quotient (ω^2) can be obtained by equating $\mathbf{U}_{\max}$ and $\mathbf{T}_{\max}$ and differentiating this quotient partially with respect to unknown constants as follows:

$$\frac{\partial \omega^2}{\partial c_i} = 0, \quad i = 1, 2, \ldots, n \tag{8.5}$$

Further manipulation of Eq. (8.5) yields the generalized eigenvalue problem of the form

$$\sum_{j=1}^{n} (a_{ij} - \lambda^2 b_{ij}) c_j = 0, \quad i = 1, 2, \ldots, n \tag{8.6}$$

where

$$a_{ij} = \c{C} \int_{\Omega} \left\{ \left(\varphi_i^{\xi\xi} \varphi_j^{\xi\xi} + \mu^4 \varphi_i^{\eta\eta} \varphi_j^{\eta\eta} \right) + 2\nu\mu^2 \left(\varphi_i^{\xi\xi} \varphi_j^{\eta\eta} + \varphi_i^{\eta\eta} \varphi_j^{\xi\xi} \right) \right.$$
$$\left. + 2(1-\nu)\mu^2 \varphi_i^{\xi\eta} \varphi_j^{\xi\eta} + K_{\mathrm{w}} \varphi_i \varphi_j + K_{\mathrm{p}} \left(\varphi_i^{\xi} \varphi_j^{\xi} + \mu^2 \varphi_i^{\eta} \varphi_j^{\eta} \right) \right\} \, \mathrm{d}\xi \, \mathrm{d}\eta$$

$$b_{ij} = \left\{ \frac{(1 - 1/\rho_{\mathrm{r}})}{k+1} + \frac{1}{\rho_{\mathrm{r}}} \right\} \int_{\Omega} \varphi_i \varphi_j \, \mathrm{d}\xi \, \mathrm{d}\eta, \quad \lambda^2 = \frac{\omega^2 a^4 \rho_{\mathrm{c}} h}{D_{\mathrm{c}}}$$

with $\c{C} = \left[12 \left(1 - \frac{1}{E_{\mathrm{r}}} \right) \left\{ \frac{1}{k+3} - \frac{1}{k+2} + \frac{1}{4(k+1)} \right\} + \frac{1}{E_{\mathrm{r}}} \right]$, $E_{\mathrm{r}} = E_{\mathrm{c}}/E_{\mathrm{m}}$, and $\rho_{\mathrm{r}} = \rho_{\mathrm{c}}/\rho_{\mathrm{m}}$ are the ratios of Young's moduli and mass densities of FG constituents, respectively, and $D_{\mathrm{c}} = E_{\mathrm{c}} h^3 / 12(1 - \nu^2)$ is meant for the flexural rigidity of the ceramic constituent of the plate. In the eigenvalue problem (Eq. (8.6)), λ and $c_j = [c_1, c_2, c_3, \ldots, c_n]^{\mathrm{T}}$ refer to the non-dimensional frequency and the column vector of unknown constants respectively.

8.1.2 Convergence and Comparison Studies

We have considered six different BCs for the FG Lévy plates, viz. SSSS, SCSS, SSSF, SCSC, SCSF, and SFSF. Here, the flow of BCs is demonstrated in Fig. 8.1. For example, if we consider the edge support SCSF, then 1 is meant for simply supported (S) and accordingly 2, 3, and 4 represent clamped (C), simply supported (S), and completely free (F) edge conditions, respectively. In Fig. 8.1 for Lévy plates, 1 and 3 are always kept simply supported and the other edges 2 and 4 vary as per different BCs.

The convergence and comparison of first six non-dimensional frequencies of isotropic Lévy plates ($k = 0$ in the case of FG plates) has been performed in Table 8.1 with or without two-parameter elastic foundations. The convergence of non-dimensional frequencies may be achieved by increasing the number of polynomials (n) in the displacement component and validation is carried out with Leissa (1973) and

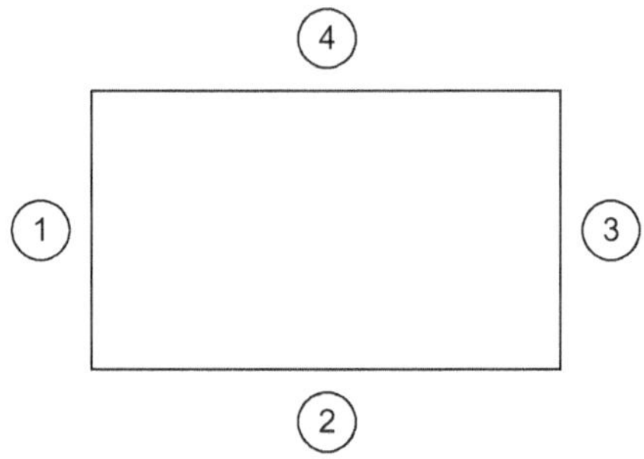

Figure 8.1 The flow of various combinations of boundary supports in FG Lévy plates. *(Reproduced from Pradhan, K.K., Chakraverty, S., Free vibration of FG Lévy plate resting on elastic foundations. Engineering and Computational Mechanics 2015;1500014 (26 pages).).*

Lam et al. (2000). From the validation, it may be noticed that present results are in excellent agreement with the available literature, irrespective of the elastic foundations and boundary support considered.

8.1.3 Results and Discussions

One may see the generalized eigenvalue problem (Eq. (8.6)), the results related to free vibration characteristics of FG Lévy plates may be influenced by four different parameters, viz. aspect ratio (μ), power-law index (k), ratio of Young's moduli (E_r), and ratio of mass densities (ρ_r). An exhaustive study of their effect on non-dimensional frequencies is carried out in the next sections. The number of polynomials involved in displacement component is taken up to 28 (as stated in Leissa, 1973), and four different cases of elastic foundations are also considered below.

(i) Case 1: $K_W = 0$ and $K_s = 0$;
(ii) Case 2: $K_W = 100$ and $K_s = 0$;
(iii) Case 3: $K_W = 0$ and $K_s = 100$;
(iv) Case 4: $K_W = 100$ and $K_s = 100$.

Effect of Aspect Ratio (μ)

The effect of aspect ratios (μ) on the first six eigenfrequencies of FG Lévy plates may be observed in Tables 8.2–8.5 with the above cases of elastic foundations, along with different combination of boundary supports. In Tables 8.2–8.5, the ascending values of μ are 0.1, 0.2, 0.5, 1.0, and 2.0, and the power-law index (k) is assumed to be unity. We have considered here FG Al/Al_2O_3 plates and hence E_r and ρ_r are also taken accordingly,

Table 8.1 Convergence and validation of the first six non-dimensional frequencies of square isotropic Lévy plates with and without elastic foundations

Sides (1234)	K_w	K_p	Source	n	λ_1	λ_2	λ_3	λ_4	λ_5	λ_6
SSSS	0	0	Present	6	19.7449	58.9915	58.9915	92.5635	139.5994	139.9505
				10	19.7449	49.5132	49.5132	92.5635	139.5994	139.9505
				15	19.7392	49.5132	49.5132	79.4007	100.1729	100.1868
				21	19.7392	49.3490	49.3490	79.4007	100.1729	100.1868
				28	19.7392	49.3490	49.3490	78.9633	98.7193	98.7194
			Leissa (1973)		19.7392	49.3480	49.3480	78.9568	98.6960	98.6980
			Lam et al. (2000)		19.74	-	-	-	-	98.69
	100	100	Present	6	49.6374	93.7017	93.7017	130.6446	175.4651	175.8185
				10	49.6374	86.5505	86.5505	130.6446	175.4651	175.8185
				15	49.6342	86.5505	86.5505	119.6471	141.6475	141.6547
				21	49.6342	86.4300	86.4300	119.6471	141.6475	141.6547
				28	49.6342	86.4300	86.4300	119.2940	140.4129	140.4129
			Lam et al. (2000)		49.63	-	-	-	-	140.39
SCSS	0	0	Present	6	23.6839	59.3808	60.8613	112.9597	143.7141	188.3806
				10	23.6484	52.4178	59.0289	92.4408	117.2825	141.0695
				15	23.6465	51.8317	58.6604	87.4429	103.5119	115.9702
				21	23.6463	51.6878	58.6505	86.2451	101.8059	113.3767
				28	23.6463	51.6754	58.6464	86.1580	100.4370	113.2943
			Leissa (1973)		23.6463	51.6743	58.6464	86.1345	100.2698	113.2281
			Lam et al. (2000)		23.65	-	-	-	-	113.23
	100	100	Present	6	52.3986	93.9500	95.4994	150.0129	179.9075	222.0115
				10	52.3095	88.9811	93.7799	130.4488	154.9408	176.9928

Continued

Table 8.1 Convergence and validation of the first six non-dimensional frequencies of square isotropic Lévy plates with and without elastic foundations—cont'd

Sides (1234)	K_w	K_p	Source	n	λ_1	λ_2	λ_3	λ_4	λ_5	λ_6
				15	52.2900	88.3913	93.6491	126.3429	144.7836	154.5166
				21	52.2868	88.2490	93.6308	125.2959	143.1058	152.6885
				28	52.2864	88.2303	93.6250	125.2078	141.8965	152.6275
			Lam et al. (2000)		52.29	-	-	-	-	152.57
SSSF	0	0	Present	6	11.7425	29.7563	51.9222	69.7889	81.2621	133.4299
				10	11.6932	28.1215	41.5901	68.6699	76.9290	118.6152
				15	11.6851	27.8133	41.4342	60.0336	63.5644	92.3070
				21	11.6846	27.7575	41.2203	59.3603	62.4616	92.0709
				28	11.6845	27.7565	41.2019	59.1001	61.8756	90.4624
			Leissa (1973)		11.6845	27.7563	41.1967	59.0655	61.8606	90.2941
			Lam et al. (2000)		11.68	-	-	-	-	90.29
	100	100	Present	6	38.5174	67.2652	85.2154	111.2268	135.6676	168.5720
				10	38.4856	65.7311	77.8585	107.3175	122.1629	168.4878
				15	38.4767	65.0341	77.7426	100.5745	111.4925	133.6523
				21	38.4746	64.9238	77.6025	99.6789	108.4398	133.4568
				28	38.4742	64.9204	77.5879	99.4255	107.6855	132.1688
			Lam et al. (2000)		38.47	-	-	-	-	132.05

Source: Reproduced from Pradhan, K.K., Chakraverty, S., Free vibration of FG Lévy plate resting on elastic foundations. Engineering and Computational Mechanics 2015;1500014 (26 pages).

Table 8.2 Effect of μ on the first six non-dimensional frequencies of functionally graded Lévy plates; $K_W = 0, K_S = 0, k = 1$

BCs	μ	λ_1	λ_2	λ_3	λ_4	λ_5	λ_6
SSSS	0.1	8.2941	8.5406	8.9518	9.6184	10.4612	14.5537
	0.2	8.5405	9.5260	11.1694	13.6727	17.0187	33.1773
	0.5	10.2650	16.4242	26.6937	34.9018	41.0638	42.3643
	1.0	16.4240	41.0608	41.0608	65.7015	82.1395	82.1395
	2.0	41.0600	65.6969	106.7748	139.6073	164.2550	169.4571
SCSS	0.1	8.2984	8.5572	8.9876	9.5946	10.6647	12.1939
	0.2	8.5717	9.6444	11.4175	13.8934	17.8138	23.1721
	0.5	10.7489	17.9171	29.3106	35.1491	41.9858	44.9437
	1.0	19.6749	42.9966	48.7968	71.6878	83.5686	94.2665
	2.0	57.6836	78.7001	116.7609	173.3926	176.9454	195.2297
SSSF	0.1	8.2109	8.3773	8.7046	9.2544	10.2004	13.0633
	0.2	8.2420	8.8839	10.1516	12.1819	15.5865	27.2149
	0.5	8.5695	12.2867	19.6585	31.6011	33.1123	37.0897
	1.0	9.7221	23.0948	34.2820	49.1742	51.4836	75.2693
	2.0	13.4249	38.8897	62.6398	79.9821	92.4091	138.5145
SCSC	0.1	8.3022	8.5737	9.0214	9.6588	10.4828	12.4449
	0.2	8.6072	9.7772	11.6861	14.3366	17.7185	24.5630
	0.5	11.3872	19.6749	32.1956	35.4375	42.9978	49.0252
	1.0	24.0886	45.5497	57.6836	78.7005	85.0726	107.4157
	2.0	79.2632	96.3545	130.1092	186.8605	211.4559	230.7359
SCSF	0.1	8.2073	8.3858	8.7229	9.2483	10.4075	11.7653
	0.2	8.2458	8.9473	10.3151	12.3524	15.8423	20.7346
	0.5	8.6745	13.1081	21.4643	33.1526	33.8266	37.5752
	1.0	10.5565	27.5118	34.7022	52.4568	60.2561	75.5222
	2.0	18.9836	42.2273	82.2810	83.0180	110.0703	144.9403
SFSF	0.1	8.1935	8.2673	8.5023	9.0049	9.6974	12.8216
	0.2	8.1651	8.4425	9.3581	11.1318	13.6167	24.1242
	0.5	8.1011	9.7248	14.7165	23.8618	32.6761	34.3452
	1.0	8.0138	13.4254	30.5587	32.4266	38.9103	59.9520
	2.0	7.9148	22.8999	32.0601	53.7084	72.7003	87.7749

Source: Reproduced from Pradhan, K.K., Chakraverty, S., Free vibration of FG Lévy plate resting on elastic foundations. Engineering and Computational Mechanics 2015;1500014 (26 pages).

which will remain the same for Tables 8.6–8.9. The important results in Tables 8.2–8.5 are summarized as below:

- $\mu = 1$ provides the eigenfrequencies of square FG plate and the remaining values of μ are for rectangular ones.
- Irrespective of the elastic foundations and edge support, it can easily be noticed that the non-dimensional frequencies follow the ascending

Table 8.3 Effect of μ on the first six non-dimensional frequencies of functionally graded Lévy plates; $K_W = 100$, $K_S = 0$, $k = 1$

BCs	μ	λ_1	λ_2	λ_3	λ_4	λ_5	λ_6
SSSS	0.1	13.6278	13.7791	14.0377	14.4719	15.0453	18.1310
	0.2	13.7791	14.4107	15.5460	17.4317	20.1633	34.8949
	0.5	14.9095	19.6641	28.8006	36.5385	42.4636	43.7225
	1.0	19.6640	42.4607	42.4607	66.5853	82.8481	82.8482
	2.0	42.4600	66.5808	107.3209	140.0255	164.6106	169.8018
SCSS	0.1	13.6303	13.7894	14.0606	14.4561	15.1875	16.2977
	0.2	13.7985	14.4892	15.7252	17.6054	20.8388	25.5709
	0.5	15.2467	20.9271	31.2416	36.7748	43.3558	46.2261
	1.0	22.4505	44.3354	49.9805	72.4987	84.2653	94.8847
	2.0	58.6883	79.4394	117.2605	173.7294	177.2755	195.5289
SSSF	0.1	13.5772	13.6785	13.8814	14.2326	14.8651	16.9580
	0.2	13.5961	13.9945	14.8317	16.2887	18.9700	29.2844
	0.5	13.7971	16.3672	22.4361	33.3999	34.8332	38.6338
	1.0	14.5411	25.5008	35.9469	50.3490	52.6069	76.0420
	2.0	17.2381	40.3650	63.5663	80.7097	93.0396	138.9359
SCSC	0.1	13.6327	13.7997	14.0822	14.4988	15.0603	16.4863
	0.2	13.8205	14.5780	15.9213	17.9572	20.7574	26.8377
	0.5	15.7033	22.4505	33.9629	37.0505	44.3366	50.2035
	1.0	26.4042	46.8156	58.6884	79.4399	85.7571	107.9586
	2.0	79.9974	96.9594	130.5577	187.1731	211.7322	230.9892
SCSF	0.1	13.5751	13.6838	13.8929	14.2286	15.0080	15.9795
	0.2	13.5984	14.0349	14.9441	16.4166	19.1807	23.3847
	0.5	13.8626	16.9925	24.0341	34.8715	35.5129	39.1001
	1.0	15.1117	29.5605	36.3478	53.5597	61.2187	76.2924
	2.0	21.8472	43.5898	82.9885	83.7192	110.6002	145.3431
SFSF	0.1	13.5668	13.6114	13.7554	14.0717	14.5246	16.7725
	0.2	13.5496	13.7186	14.3002	15.5190	17.3878	26.4368
	0.5	13.5111	14.5429	18.2619	26.1975	34.4188	36.0072
	1.0	13.4590	17.2384	32.4154	34.1820	40.3848	60.9193
	2.0	13.4003	25.3245	33.8345	54.7861	73.5000	88.4384

Source: Reproduced from Pradhan, K.K., Chakraverty, S., Free vibration of FG Lévy plate resting on elastic foundations. Engineering and Computational Mechanics 2015;1500014 (26 pages).

behavior with an increase in μ except for slight fluctuations on fundamental frequencies in the case of SFSF.

- It is also evident that the two-parameter elastic foundations (Case 4. effect of both Winkler and Pasternak foundations) yield higher values for the non-dimensional frequencies than other three cases. The least values for the frequencies are obtained for the FG plates without elastic foundations (Case 1).

Table 8.4 Effect of μ on the first six non-dimensional frequencies of functionally graded Lévy plates; $K_W = 0$, $K_S = 100$, $k = 1$

BCs	μ	λ_1	λ_2	λ_3	λ_4	λ_5	λ_6
SSSS	0.1	35.1328	35.6803	36.5786	37.8866	39.5476	45.2362
	0.2	35.6803	37.8070	41.1607	45.7414	51.3987	71.0557
	0.5	39.3427	50.7713	66.8070	78.2470	86.3498	87.4718
	1.0	50.7712	86.3478	86.3478	116.3991	135.2312	135.2312
	2.0	86.3473	116.3958	162.4913	197.7575	223.7408	228.3980
SCSS	0.1	35.1382	35.7013	36.6242	37.8908	39.6978	41.9155
	0.2	35.7068	37.9072	41.3721	45.9035	51.8321	58.9640
	0.5	39.6641	51.8029	68.6912	78.4417	87.0805	89.2395
	1.0	52.9117	87.7742	92.0272	121.0661	136.4214	144.8406
	2.0	98.4973	126.3626	170.6411	226.2614	234.9829	250.3040
SSSF	0.1	34.9945	35.3613	36.0875	37.2107	39.0190	43.3038
	0.2	35.1312	36.5756	39.3390	43.3657	49.7031	64.0780
	0.5	36.0779	44.3190	58.1615	76.1261	76.5798	81.6040
	1.0	39.3039	66.3039	78.1248	98.6982	107.1168	128.0310
	2.0	50.4358	85.8271	128.8996	134.4314	156.6912	201.4463
SCSC	0.1	35.1413	35.7193	36.6508	37.9612	39.5691	42.0653
	0.2	35.7285	38.0023	41.5395	46.2400	51.7812	59.5769
	0.5	40.0155	52.9203	70.6625	78.6446	87.7875	92.2285
	1.0	55.4720	89.4729	98.4977	126.3643	137.5274	155.3614
	2.0	114.2046	139.4251	181.0151	243.2219	259.2071	281.1895
SCSF	0.1	34.9958	35.3768	36.1243	37.2331	38.9625	41.5248
	0.2	35.1379	36.6436	39.5053	43.5265	48.9661	58.1707
	0.5	36.1647	44.9787	59.6039	76.1796	78.1140	82.0454
	1.0	39.9676	70.0250	78.4995	101.4933	114.2805	128.3487
	2.0	54.9188	88.7469	136.5755	146.5920	172.6097	203.6114
SFSF	0.1	34.9485	35.1323	35.6796	36.6830	38.0944	46.8325
	0.2	34.9474	35.6776	37.8003	41.4416	46.4805	72.0597
	0.5	34.9429	39.3115	50.6117	68.1610	75.4416	78.1582
	1.0	34.9359	50.4453	75.3911	84.2015	85.8584	116.1225
	2.0	34.9230	75.2935	82.2905	112.3440	125.4598	160.0090

Source: Reproduced from Pradhan, K.K., Chakraverty, S., Free vibration of FG Lévy plate resting on elastic foundations. Engineering and Computational Mechanics 2015;1500014 (26 pages).

Effect of Power-Law Index (k)

The variation of the first six non-dimensional frequencies of rectangular FG Al/Al_2O_3 plates with increasing values of power-law index ($k = 0.1$, 0.2, 0.5, 1.0, 2.0, and 5.0) is reported in Tables 8.6–8.9 subject to various edge supports. Looking into these computations, we may draw the following conclusions.

Table 8.5 Effect of μ on the first six non-dimensional frequencies of functionally graded Lévy plates; $K_W = 100$, $K_S = 100$, $k = 1$

BCs	μ	λ_1	λ_2	λ_3	λ_4	λ_5	λ_6
SSSS	0.1	36.7592	37.2828	38.1434	39.3995	40.9993	46.5106
	0.2	37.2828	39.3229	42.5574	47.0021	52.5238	71.8737
	0.5	40.8016	51.9100	67.6764	78.9906	87.0242	88.1376
	1.0	51.9099	87.0222	87.0222	116.9003	135.6629	135.6629
	2.0	87.0217	116.8970	162.8507	198.0529	224.0019	228.6538
SCSS	0.1	36.7643	37.3029	38.1871	39.4035	41.1441	43.2878
	0.2	37.3082	39.4193	42.7618	47.1599	52.9480	59.9473
	0.5	41.1116	52.9194	69.5371	79.1835	87.7493	89.8922
	1.0	54.0053	88.4378	92.6603	121.5480	136.8493	145.2436
	2.0	99.0890	126.8244	170.9834	226.5196	235.2316	250.5374
SSSF	0.1	36.6270	36.9776	37.6727	38.7499	40.4896	44.6334
	0.2	36.7576	38.1405	40.7980	44.6935	50.8657	64.9840
	0.5	37.6635	45.6191	59.1581	76.8902	77.3395	82.3173
	1.0	40.7642	67.1798	78.8695	99.2887	107.6612	128.4868
	2.0	51.5819	86.5055	129.3523	134.8655	157.0638	201.7363
SCSC	0.1	36.7673	37.3201	38.2126	39.4712	41.0200	43.4329
	0.2	37.3289	39.5107	42.9238	47.4875	52.8982	60.5503
	0.5	41.4507	54.0137	71.4851	79.3845	88.4510	92.8602
	1.0	56.5161	90.1240	99.0895	126.8261	137.9518	155.7373
	2.0	114.7153	139.8438	181.3378	243.4621	259.4326	281.3973
SCSF	0.1	36.6283	36.9924	37.7080	38.7715	40.4351	42.9096
	0.2	36.7640	38.2057	40.9584	44.8495	50.1458	59.1672
	0.5	37.7466	46.2602	60.5768	76.9432	78.8589	82.7548
	1.0	41.4045	70.8549	79.2407	102.0677	114.7909	128.8034
	2.0	55.9732	89.4033	137.0029	146.9903	172.9481	203.8983
SFSF	0.1	36.5830	36.7587	37.2821	38.2436	39.5993	48.0646
	0.2	36.5820	37.2802	39.3165	42.8290	47.7217	72.8665
	0.5	36.5778	40.7716	51.7539	69.0134	76.2126	78.9026
	1.0	36.5711	51.5912	76.1626	84.8929	86.5366	116.6249
	2.0	36.5587	76.0659	82.9979	112.8632	125.9249	160.3740

Source: Reproduced from Pradhan, K.K., Chakraverty, S., Free vibration of FG Lévy plate resting on elastic foundations. Engineering and Computational Mechanics 2015;1500014 (26 pages).

- Case 1: The non-dimensional frequencies follow a descending pattern with the ascending values of k in the absence of elastic foundations.
- Case 2: In the presence of the Winkler foundation, the eigenfrequencies are decreasing with the increasing values of k except for the SFSF FG plates. In the SFSF FG plates, we may get fluctuations in the fundamental frequencies with an ascending value of k.

Table 8.6 Effect of k on the first six non-dimensional frequencies of functionally graded Lévy plates; $K_W = 0, K_S = 0, \mu = 2.0$

BCs	k	λ_1	λ_2	λ_3	λ_4	λ_5	λ_6
SSSS	0.0	49.3480	78.9579	128.3273	167.7871	197.4100	203.6621
	0.1	47.5844	76.1360	123.7410	161.7906	190.3548	196.3835
	0.2	46.1705	73.8738	120.0643	156.9834	184.6988	190.5484
	0.5	43.3429	69.3495	112.7111	147.3691	173.3872	178.8785
	1.0	41.0600	65.6969	106.7748	139.6073	164.2550	169.4571
	2.0	39.2498	62.8005	102.0673	133.4523	157.0134	161.9861
	5.0	36.5961	58.5545	95.1665	124.4296	146.3977	151.0342
SCSS	0.0	69.3270	94.5857	140.3291	208.3919	212.6619	234.6369
	0.1	66.8494	91.2053	135.3139	200.9443	205.0617	226.2512
	0.2	64.8631	88.4954	131.2934	194.9736	198.9687	219.5286
	0.5	60.8906	83.0756	123.2525	183.0327	186.7831	206.0839
	1.0	57.6836	78.7001	116.7609	173.3926	176.9454	195.2297
	2.0	55.1404	75.2304	111.6132	165.7481	169.1443	186.6224
	5.0	51.4124	70.1440	104.0670	154.5418	157.7084	174.0048
SSSF	0.0	16.1348	46.7396	75.2837	96.1266	111.0619	166.4737
	0.1	15.5581	45.0692	72.5932	92.6911	107.0927	160.5241
	0.2	15.0959	43.7300	70.4362	89.9370	103.9106	155.7545
	0.5	14.1713	41.0518	66.1224	84.4289	97.5468	146.2155
	1.0	13.4249	38.8897	62.6398	79.9821	92.4091	138.5145
	2.0	12.8331	37.1751	59.8782	76.4559	88.3350	132.4077
	5.0	11.9654	34.6617	55.8298	71.2867	82.3626	123.4555
SCSC	0.0	95.2625	115.8037	156.3718	224.5784	254.1383	277.3101
	0.1	91.8580	111.6651	150.7833	216.5523	245.0558	267.3994
	0.2	89.1286	108.3472	146.3030	210.1179	237.7744	259.4542
	0.5	83.6700	101.7116	137.3429	197.2495	223.2122	243.5642
	1.0	79.2632	96.3545	130.1092	186.8605	211.4559	230.7359
	2.0	75.7687	92.1065	124.3730	178.6223	202.1332	220.5633
	5.0	70.6459	85.8791	115.9641	166.5455	188.4669	205.6509
SCSF	0.0	22.8155	50.7509	98.8895	99.7752	132.2880	174.1965
	0.1	22.0001	48.9371	95.3553	96.2094	127.5602	167.9710
	0.2	21.3464	47.4831	92.5220	93.3507	123.7701	162.9801
	0.5	20.0390	44.5750	86.8556	87.6336	116.1899	152.9986
	1.0	18.9836	42.2273	82.2810	83.0180	110.0703	144.9403
	2.0	18.1467	40.3656	78.6535	79.3579	105.2175	138.5502
	5.0	16.9198	37.6365	73.3357	73.9925	98.1038	129.1828
SFSF	0.0	9.5125	27.5223	38.5314	64.5495	87.3748	105.4923
	0.1	9.1725	26.5386	37.1544	62.2426	84.2522	101.7222
	0.2	8.9000	25.7501	36.0504	60.3931	81.7488	98.6997
	0.5	8.3549	24.1731	33.8425	56.6944	76.7422	92.6550
	1.0	7.9148	22.8999	32.0601	53.7084	72.7003	87.7749
	2.0	7.5659	21.8903	30.6466	51.3405	69.4951	83.9051
	5.0	7.0544	20.4103	28.5746	47.8694	64.7965	78.2323

Source: Reproduced from Pradhan, K.K., Chakraverty, S., Free vibration of FG Lévy plate resting on elastic foundations. Engineering and Computational Mechanics 2015;1500014 (26 pages).

Table 8.7 Effect of k on the first six non-dimensional frequencies of functionally graded Lévy plates; $K_W = 100, K_S = 0, \mu = 2.0$

BCs	k	λ_1	λ_2	λ_3	λ_4	λ_5	λ_6
SSSS	0.0	50.3510	79.5886	128.7163	168.0849	197.6631	203.9075
	0.1	48.6516	76.8075	124.1553	162.1077	190.6244	196.6448
	0.2	47.2947	74.5815	120.5011	157.3176	184.9830	190.8239
	0.5	44.6014	70.1429	113.2011	147.7442	173.7060	179.1876
	1.0	42.4600	66.5808	107.3209	140.0255	164.6106	169.8018
	2.0	40.7978	63.7794	102.6725	133.9158	157.4075	162.3682
	5.0	38.3545	59.6693	95.8564	124.9581	146.8471	151.4699
SCSS	0.0	70.0445	95.1129	140.6850	208.6317	212.8969	234.8499
	0.1	67.6132	91.7666	135.6929	201.1997	205.3119	226.4781
	0.2	65.6680	89.0870	131.6929	195.2429	199.2325	219.7678
	0.5	61.7928	83.7391	123.7007	183.3348	187.0791	206.3523
	1.0	58.6883	79.4394	117.2605	173.7294	177.2755	195.5289
	2.0	56.2528	76.0494	112.1669	166.1214	169.5102	186.9541
	5.0	52.6785	71.0772	104.6982	154.9676	158.1257	174.3831
SSSF	0.0	18.9824	47.7974	75.9449	96.6453	111.5112	166.7737
	0.1	18.5677	46.1945	73.2971	93.2435	107.5711	160.8437
	0.2	18.2470	44.9153	71.1781	90.5192	104.4150	156.0914
	0.5	17.6495	42.3784	66.9541	85.0819	98.1124	146.5935
	1.0	17.2381	40.3650	63.5663	80.7097	93.0396	138.9359
	2.0	16.9883	38.8060	60.9041	77.2620	89.0336	132.8748
	5.0	16.5820	36.5133	56.9979	72.2051	83.1588	123.9882
SCSC	0.0	95.7859	116.2347	156.6912	224.8009	254.3350	277.4903
	0.1	92.4153	112.1240	151.1234	216.7893	245.2652	267.5914
	0.2	89.7161	108.8310	146.6617	210.3677	237.9953	259.6566
	0.5	84.3288	102.2542	137.7452	197.5298	223.4600	243.7913
	1.0	79.9974	96.9594	130.5577	187.1731	211.7322	230.9892
	2.0	76.5820	92.7767	124.8701	178.9688	202.4395	220.8440
	5.0	71.5726	86.6430	116.5309	166.9407	188.8163	205.9711
SCSF	0.0	24.9107	51.7267	99.3938	100.2751	132.6655	174.4833
	0.1	24.2220	49.9755	95.8923	96.7416	127.9622	168.2764
	0.2	23.6799	48.5769	93.0881	93.9118	124.1938	163.3021
	0.5	22.6328	45.7997	87.4905	88.2628	116.6652	153.3599
	1.0	21.8472	43.5898	82.9885	83.7192	110.6002	145.3431
	2.0	21.2888	41.8724	79.4373	80.1348	105.8048	138.9966
	5.0	20.4468	39.3484	74.2288	74.8778	98.7732	129.6919
SFSF	0.0	13.8017	29.2827	39.8079	65.3195	87.9452	105.9652
	0.1	13.6688	28.4078	38.5117	63.0622	84.8595	102.2257
	0.2	13.5749	27.7153	37.4793	61.2568	82.3889	99.2305
	0.5	13.4344	26.3632	35.4401	57.6623	77.4600	93.2503
	1.0	13.4003	25.3245	33.8345	54.7861	73.5000	88.4384
	2.0	13.4594	24.5581	32.6057	52.5334	70.3809	84.6403
	5.0	13.4743	23.4173	30.7945	49.2267	65.8056	79.0701

Source: Reproduced from Pradhan, K.K., Chakraverty, S., Free vibration of FG Lévy plate resting on elastic foundations. Engineering and Computational Mechanics 2015;1500014 (26 pages).

Table 8.8 Effect of k on the first six non-dimensional frequencies of functionally graded Lévy plates; $K_W = 0, K_S = 100, \mu = 2.0$

BCs	k	λ_1	λ_2	λ_3	λ_4	λ_5	λ_6
SSSS	0.0	85.8489	118.8699	171.1696	211.9691	242.3022	248.0679
	0.1	85.6297	117.9228	168.7892	208.3461	237.7144	243.2437
	0.2	85.5375	117.2744	167.0240	205.6032	234.2092	239.5492
	0.5	85.6763	116.3971	164.0282	200.7185	227.8396	232.8018
	1.0	86.3473	116.3958	162.4913	197.7575	223.7408	228.3980
	2.0	87.4952	117.1653	162.2257	196.4692	221.6150	226.0303
	5.0	88.5604	117.6203	161.1430	193.8951	217.8246	221.8982
SCSS	0.0	101.4516	131.6475	181.5286	248.0430	256.2871	275.8023
	0.1	100.4857	130.0973	178.6782	242.8202	251.1163	269.7513
	0.2	99.8003	128.9680	176.5367	238.7928	247.1438	265.0702
	0.5	98.7658	127.1346	172.7887	231.3288	239.8394	256.3366
	1.0	98.4973	126.3626	170.6411	226.2614	234.9829	250.3040
	2.0	98.9015	126.5184	169.8865	223.2750	232.2528	246.6168
	5.0	98.9471	126.1281	168.1235	218.3189	227.6097	240.6269
SSSF	0.0	47.9790	84.9364	129.4577	138.7482	161.8780	215.6343
	0.1	48.3180	84.8068	128.9072	137.3613	160.2552	212.2227
	0.2	48.6327	84.7839	128.5743	136.3734	159.0903	209.6513
	0.5	49.4430	85.0546	128.3358	134.8653	157.2778	204.5658
	1.0	50.4358	85.8271	128.8996	134.4314	156.6912	201.4463
	2.0	51.6086	87.0530	130.1887	134.9396	157.1841	200.0365
	5.0	52.8471	88.2111	131.1729	134.9724	157.0511	197.2775
SCSC	0.0	122.0333	148.7463	194.9879	266.7988	289.5932	314.7085
	0.1	120.0267	146.3376	191.4857	261.1501	282.5636	306.9744
	0.2	118.5097	144.5212	188.8211	256.7931	277.0815	300.9378
	0.5	115.8172	141.3144	184.0225	248.7142	266.6846	289.4700
	1.0	114.2046	139.4251	181.0151	243.2219	259.2071	281.1895
	2.0	113.5313	138.6821	179.5647	239.9760	254.2645	275.6764
	5.0	112.1092	137.0636	176.8407	234.6105	246.5411	267.0991
SCSF	0.0	53.5455	88.5609	141.3757	151.1314	181.4436	218.8095
	0.1	53.6543	88.2819	139.8862	149.7278	179.0440	214.9737
	0.2	53.7845	88.1391	138.8161	148.7142	177.2572	212.0611
	0.5	54.2256	88.1696	137.1431	147.1207	174.2024	206.8394
	1.0	54.9188	88.7469	136.5755	146.5920	172.6097	203.6114
	2.0	55.8521	89.8166	136.9770	147.0113	172.2992	202.1171
	5.0	56.7508	90.7526	136.8566	146.7416	171.0121	199.2334
SFSF	0.0	32.8799	73.8986	78.7658	112.1846	128.7988	166.6219
	0.1	33.1855	73.9114	79.2369	111.8194	127.6431	164.6832
	0.2	33.4605	73.9932	79.6785	111.6280	126.8342	163.2700
	0.5	34.1371	74.4382	80.8350	111.6409	125.6606	160.9747
	1.0	34.9230	75.2935	82.2905	112.3440	125.4598	160.0090
	2.0	35.8198	76.5224	84.0428	113.6687	126.1135	160.2074
	5.0	36.7843	77.7333	85.8309	114.8076	126.3762	159.7007

Source: Reproduced from Pradhan, K.K., Chakraverty, S., Free vibration of FG Lévy plate resting on elastic foundations. Engineering and Computational Mechanics 2015;1500014 (26 pages).

Table 8.9 Effect of k on the first six non-dimensional frequencies of functionally graded Lévy plates; $K_W = 100$, $K_S = 100$, $\mu = 2.0$

BCs	k	λ_1	λ_2	λ_3	λ_4	λ_5	λ_6
SSSS	0.0	86.4293	119.2898	171.4615	212.2049	242.5085	248.2694
	0.1	86.2273	118.3575	169.0931	208.5924	237.9303	243.4547
	0.2	86.1495	117.7215	167.3382	205.8586	234.4334	239.7684
	0.5	86.3198	116.8716	164.3653	200.9940	228.0824	233.0394
	1.0	87.0217	116.8970	162.8507	198.0529	224.0019	228.6538
	2.0	88.2005	117.6929	162.6072	196.7843	221.8944	226.3042
	5.0	89.3014	118.1792	161.5514	194.2346	218.1269	222.1950
SCSS	0.0	101.9432	132.0267	181.8038	248.2445	256.4821	275.9835
	0.1	100.9955	130.4915	178.9653	243.0316	251.3207	269.9416
	0.2	100.3253	129.3747	176.8340	239.0127	247.3563	265.2683
	0.5	99.3245	127.5691	173.1087	231.5679	240.0700	256.5524
	1.0	99.0890	126.8244	170.9834	226.5196	235.2316	250.5374
	2.0	99.5260	127.0071	170.2508	223.5523	232.5195	246.8679
	5.0	99.6108	126.6495	168.5149	218.6206	227.8990	240.9006
SSSF	0.0	49.0100	85.5230	129.8433	139.1081	162.1865	215.8660
	0.1	49.3693	85.4102	129.3050	137.7347	160.5753	212.4645
	0.2	49.7012	85.4013	128.9822	136.7581	159.4202	209.9017
	0.5	50.5498	85.7028	128.7663	135.2750	157.6293	204.8361
	1.0	51.5819	86.5055	129.3523	134.8655	157.0638	201.7363
	2.0	52.7954	87.7618	130.6637	135.3979	157.5778	200.3460
	5.0	54.0796	88.9550	131.6743	135.4597	157.4701	197.6112
SCSC	0.0	122.4424	149.0820	195.2442	266.9862	289.7658	314.8674
	0.1	120.4538	146.6881	191.7537	261.3467	282.7453	307.1416
	0.2	118.9522	144.8842	189.0991	256.9976	277.2711	301.1123
	0.5	116.2940	141.7055	184.3229	248.9366	266.8920	289.6611
	1.0	114.7153	139.8438	181.3378	243.4621	259.4326	281.3973
	2.0	114.0757	139.1281	179.9094	240.2340	254.5081	275.9010
	5.0	112.6955	137.5436	177.2129	234.8912	246.8082	267.3457
SCSF	0.0	54.4713	89.1237	141.7289	151.4618	181.7190	219.0379
	0.1	54.6030	88.8617	140.2528	150.0704	179.3305	215.2125
	0.2	54.7526	88.7331	139.1940	149.0670	177.5534	212.3087
	0.5	55.2367	88.7951	137.5461	147.4964	174.5197	207.1068
	1.0	55.9732	89.4033	137.0029	146.9903	172.9481	203.8983
	2.0	56.9506	90.5038	137.4286	147.4322	172.6584	202.4234
	5.0	57.9003	91.4759	137.3372	147.1900	171.3970	199.5638
SFSF	0.0	34.3670	74.5722	79.3981	112.6294	129.1865	166.9217
	0.1	34.6985	74.6029	79.8824	112.2777	128.0448	164.9947
	0.2	34.9953	74.6998	80.3351	112.0976	127.2477	163.5915
	0.5	35.7215	75.1779	81.5168	112.1355	126.1002	161.3181
	1.0	36.5587	76.0659	82.9979	112.8632	125.9249	160.3740
	2.0	37.5096	77.3278	84.7768	114.2125	126.6039	160.5936
	5.0	38.5341	78.5764	86.5952	115.3802	126.8966	160.1128

Source: Reproduced from Pradhan, K.K., Chakraverty, S., Free vibration of FG Lévy plate resting on elastic foundations. Engineering and Computational Mechanics 2015;1500014 (26 pages).

- Cases 3 and 4: With the effect of both foundations, the first six lowest frequencies of only SCSC FG Lévy plates follow descending behavior with an increase in k, whereas these frequencies fluctuate randomly at different modes for the remaining five combinations of boundary supports.

Effect of Ratio of Young's Moduli (E_r)

In this section, the effect of an increase in E_r on the first six lowest eigenfrequencies of FG plates subjected to different edge supports are evaluated in Tables 8.10–8.13. Ascending values of the ratios of Young's moduli (E_r) are taken as 0.1, 0.2, 0.5, 1.0, 2.0, and 5.0. The common numerical factors assumed in these computations are: $k = 1$, $\mu = 2.0$, and $\rho_r = 1.5$. In these tabulations, it is evident that the eigenfrequencies are also decreasing with an increase in E_r regardless of the elastic foundation and edge support assumed. This statement can be validated by looking into the expression of λ which is inversely proportional to E_c.

Effect of Ratio of Mass Densities (ρ_r)

The expression of λ conveys that the non-dimensional frequency is directly proportional to ρ_r. As such, the first six non-dimensional frequencies of FG Lévy plate with different values of ρ_r are incorporated in Tables 8.14–8.17. Increasing values of ρ_r are 0.1, 0.2, 0.5, 1.0, and 2.0 in these computations. Other major components supposed in these tabulations are: $k = 1$, $\mu = 2.0$, and $E_r = 3.0$. Irrespective of elastic foundations and edge conditions, it can easily be noted that the non-dimensional frequencies are following ascending behavior with increasing values of ρ_r.

Three-dimensional (3D) mode shapes corresponding to the first six eigenfrequencies have been depicted in Figs. 8.2–8.7 by considering FG Al/Al_2O_3 Lévy square plates with the parameters: $k = 1$, $K_w = 100$, and $K_s = 100$. All six combinations of boundary supports are also taken here. One may also depict the mode shapes of FG Lévy plates for different sets of parameters.

8.2 THERMAL ENVIRONMENTS

This investigation involves the exponential variation of temperature dependent material properties (as stated in Eq. (2.2)) to evaluate the effect of thermal environment on free vibration of FG plate. In addition, the material properties of FG constituents follow the exponential gradation form to vary

Table 8.10 Effect of E_r on the first six non-dimensional frequencies of functionally graded Lévy plates; $K_W = 0$, $K_S = 0$, $k = 1$, $\mu = 2.0$, and $\rho_r = 1.5$

BCs	E_r	λ_1	λ_2	λ_3	λ_4	λ_5	λ_6
SSSS	0.1	126.7774	202.8464	329.6788	431.0529	507.1554	523.2174
	0.2	93.6313	149.8120	243.4839	318.3537	374.5591	386.4217
	0.5	66.2073	105.9331	172.1691	225.1100	264.8533	273.2414
	1.0	54.0580	86.4940	140.5755	183.8016	216.2518	223.1007
	2.0	46.8156	74.9060	121.7420	159.1768	187.2795	193.2108
	5.0	41.8732	66.9980	108.8893	142.3721	167.5079	172.8130
SCSS	0.1	178.1043	242.9951	360.5120	535.3685	546.3384	602.7930
	0.2	131.5388	179.4638	266.2558	395.3959	403.4976	445.1921
	0.5	93.0119	126.9000	188.2713	279.5871	285.3159	314.7984
	1.0	75.9439	103.6134	153.7229	228.2819	232.9595	257.0318
	2.0	65.7694	89.7319	133.1279	197.6979	201.7488	222.5961
	5.0	58.8259	80.2586	119.0732	176.8264	180.4496	199.0960
SSSF	0.1	41.4510	120.0761	193.4073	246.9536	285.3232	427.6786
	0.2	30.6136	88.6821	142.8408	182.3873	210.7251	315.8616
	0.5	21.6471	62.7077	101.0037	128.9673	149.0052	223.3479
	1.0	17.6748	51.2006	82.4691	105.3014	121.6622	182.3628
	2.0	15.3068	44.3410	71.4204	91.1937	105.3626	157.9308
	5.0	13.6908	39.6598	63.8803	81.5661	94.2391	141.2576
SCSC	0.1	244.7338	297.5052	401.7264	576.9524	652.8932	712.4225
	0.2	180.7479	219.7221	296.6946	426.1076	482.1936	526.1589
	0.5	127.8081	155.3670	209.7948	301.3035	340.9623	372.0505
	1.0	104.3548	126.8566	171.2967	246.0133	278.3946	303.7780
	2.0	90.3739	109.8611	148.3473	213.0538	241.0968	263.0795
	5.0	80.8329	98.2627	132.6859	190.5611	215.6435	235.3054
SCSF	0.1	58.6140	130.3815	254.0517	256.3271	339.8541	447.5190
	0.2	43.2893	96.2931	187.6296	189.3101	250.9989	330.5147
	0.5	30.6102	68.0895	132.6742	133.8625	177.4830	233.7092
	1.0	24.9931	55.5948	108.3280	109.2982	144.9143	190.8228
	2.0	21.6446	48.1465	93.8148	94.6551	125.4995	165.2573
	5.0	19.3596	43.0636	83.9105	84.6621	112.2501	147.8107
SFSF	0.1	58.6140	130.3815	254.0517	256.3271	339.8541	447.5190
	0.2	43.2893	96.2931	187.6296	189.3101	250.9989	330.5147
	0.5	30.6102	68.0895	132.6742	133.8625	177.4830	233.7092
	1.0	24.9931	55.5948	108.3280	109.2982	144.9143	190.8228
	2.0	21.6446	48.1465	93.8148	94.6551	125.4995	165.2573
	5.0	19.3596	43.0636	83.9105	84.6621	112.2501	147.8107

Source: Reproduced from Pradhan, K.K., Chakraverty, S., Free vibration of FG Lévy plate resting on elastic foundations. Engineering and Computational Mechanics 2015;1500014 (26 pages).

Table 8.11 Effect of E_r on the first six non-dimensional frequencies of functionally graded Lévy plates; $K_W = 100$, $K_S = 0$, $k = 1$, $\mu = 2.0$, and $\rho_r = 1.5$

BCs	E_r	λ_1	λ_2	λ_3	λ_4	λ_5	λ_6
SSSS	0.1	127.2498	203.1420	329.8607	431.1921	507.2737	523.3321
	0.2	94.2699	150.2120	243.7302	318.5421	374.7193	386.5769
	0.5	67.1074	106.4980	172.5173	225.3764	265.0797	273.4609
	1.0	55.1568	87.1849	141.0017	184.1277	216.5291	223.3694
	2.0	48.0802	75.7028	122.2338	159.5533	187.5997	193.5211
	5.0	43.2824	67.8876	109.4389	142.7929	167.8657	173.1599
SCSS	0.1	178.4409	243.2419	360.6784	535.4806	546.4482	602.8925
	0.2	131.9941	179.7978	266.4810	395.5476	403.6463	445.3269
	0.5	93.6548	127.3720	188.5897	279.8016	285.5261	314.9889
	1.0	76.7299	104.1909	154.1127	228.5446	233.2169	257.2651
	2.0	66.6754	90.3981	133.5778	198.0012	202.0460	222.8654
	5.0	59.8372	81.0028	119.5760	177.1654	180.7818	199.3971
SSSF	0.1	42.8741	120.5748	193.7173	247.1964	285.5334	427.8189
	0.2	32.5145	89.3561	143.2602	182.7160	211.0097	316.0515
	0.5	24.2610	63.6573	101.5960	129.4317	149.4073	223.6163
	1.0	20.7942	52.3594	83.1935	105.8696	122.1544	182.6915
	2.0	18.8228	45.6742	72.2556	91.8492	105.9305	158.3102
	5.0	17.5339	41.1449	64.8128	82.2984	94.8737	141.6817
SCSC	0.1	244.9788	297.7068	401.8757	577.0563	652.9851	712.5067
	0.2	181.0795	219.9950	296.8967	426.2483	482.3180	526.2729
	0.5	128.2767	155.7527	210.0806	301.5026	341.1383	372.2118
	1.0	104.9282	127.3287	171.6466	246.2571	278.6100	303.9754
	2.0	91.0354	110.4059	148.7512	213.3352	241.3455	263.3074
	5.0	81.5718	98.8715	133.1373	190.8757	215.9216	235.5603
SCSF	0.1	59.6288	130.8408	254.2878	256.5611	340.0306	447.6531
	0.2	44.6538	96.9142	187.9492	189.6268	251.2378	330.6962
	0.5	32.5113	68.9651	133.1257	134.3099	177.8208	233.9658
	1.0	27.2884	56.6638	108.8805	109.8458	145.3277	191.1369
	2.0	24.2588	49.3770	94.4522	95.2868	125.9766	165.6200
	5.0	22.2439	44.4350	84.6226	85.3678	112.7834	148.2160
SFSF	0.1	26.7808	71.5495	99.5933	166.1920	224.7372	271.2360
	0.2	21.1129	53.3564	73.9244	122.9629	166.1436	200.4571
	0.5	16.8189	38.5156	52.8432	87.2923	117.7364	141.9561
	1.0	15.1190	32.0776	43.6074	71.5539	96.3392	116.0791
	2.0	14.1929	28.3148	38.1602	62.2091	83.6118	100.6765
	5.0	13.6070	25.7950	34.4814	55.8567	74.9449	90.1810

Source: Reproduced from Pradhan, K.K., Chakraverty, S., Free vibration of FG Lévy plate resting on elastic foundations. Engineering and Computational Mechanics 2015;1500014 (26 pages).

Table 8.12 Effect of E_r on the first six non-dimensional frequencies of functionally graded Lévy plates; $K_W = 0$, $K_S = 100$, $k = 1$, $\mu = 2.0$, and $\rho_r = 1.5$

BCs	E_r	λ_1	λ_2	λ_3	λ_4	λ_5	λ_6
SSSS	0.1	148.3046	224.9923	352.2577	453.8069	529.9946	545.7410
	0.2	121.1965	178.6575	273.2797	348.5443	404.9471	416.4102
	0.5	101.5144	143.8634	212.2253	266.0989	306.3248	314.2202
	1.0	94.0427	130.2154	187.5069	232.2006	265.4288	271.7448
	2.0	90.0748	122.8241	173.8346	213.2401	242.4072	247.7895
	5.0	87.6078	118.1676	165.0886	201.0072	227.4786	232.2316
SCSS	0.1	195.8818	262.9715	381.9812	555.3139	568.3775	623.3883
	0.2	154.7164	205.6821	294.6645	422.0020	432.8674	472.6970
	0.5	123.5256	161.8095	226.6480	316.0729	325.5175	352.6027
	1.0	111.1346	144.2126	198.8546	271.7175	280.7484	302.1262
	2.0	104.3448	134.5281	183.3728	246.5493	255.4322	273.4016
	5.0	100.0188	128.3486	173.4123	230.1171	238.9537	254.6044
SSSF	0.1	65.2474	143.2746	226.4470	270.3026	313.7477	453.5410
	0.2	58.7112	118.1245	184.7873	212.9012	247.6833	350.0076
	0.5	54.2235	99.9458	153.9574	169.3122	197.4346	269.2877
	1.0	52.5583	93.0431	141.8138	151.9911	177.3284	236.2155
	2.0	51.6753	89.3661	135.1222	142.5219	166.2347	217.3742
	5.0	51.1246	87.0718	130.8186	136.5127	159.1249	204.7653
SCSC	0.1	258.7179	314.6684	421.5475	598.1576	670.4406	730.8851
	0.2	199.2441	242.4356	323.0150	454.4025	505.6825	550.8858
	0.5	152.7695	186.0532	245.5806	340.1298	373.3945	406.2282
	1.0	133.6808	162.9434	213.5986	292.2635	317.2334	344.7459
	2.0	122.9905	150.0321	195.6408	265.1012	285.0051	309.4343
	5.0	116.0740	141.7006	184.0142	247.3701	263.7620	286.1445
SCSF	0.1	80.7628	153.2430	277.3079	286.3158	366.8961	470.5321
	0.2	69.8605	125.3532	218.0289	227.9098	286.3263	361.0374
	0.5	61.8571	104.9130	172.8815	183.5186	224.1118	275.1201
	1.0	58.6562	97.0137	154.8693	165.5561	198.7615	239.6938
	2.0	56.8652	92.7463	144.9891	155.4657	184.5732	219.8247
	5.0	55.7004	90.0521	138.7004	148.8699	175.3683	206.9722
SFSF	0.1	42.3159	108.6547	120.7989	194.4692	247.3163	312.9922
	0.2	38.9570	97.2947	100.6592	158.9574	195.5870	253.8024
	0.5	36.7798	86.3155	89.3660	132.9477	156.5195	204.6414
	1.0	36.0181	80.9519	86.2836	122.8921	141.0921	182.5251
	2.0	35.6288	78.1232	84.5595	117.4475	132.6958	170.3267
	5.0	35.3921	76.3712	83.4284	113.9987	127.3857	162.5233

Source: Reproduced from Pradhan, K.K., Chakraverty, S., Free vibration of FG Lévy plate resting on elastic foundations. Engineering and Computational Mechanics 2015;1500014 (26 pages).

Table 8.13 Effect of E_r on the first six non-dimensional frequencies of functionally graded Lévy plates; $K_W = 100$, $K_S = 100$, $k = 1$, $\mu = 2.0$, and $\rho_r = 1.5$

BCs	E_r	λ_1	λ_2	λ_3	λ_4	λ_5	λ_6
SSSS	0.1	148.7087	225.2588	352.4280	453.9391	530.1078	545.8510
	0.2	121.6905	178.9930	273.4992	348.7164	405.0952	416.5543
	0.5	102.1037	144.2798	212.5079	266.3243	306.5206	314.4111
	1.0	94.6786	130.6754	187.8267	232.4588	265.6548	271.9655
	2.0	90.7385	123.3116	174.1794	213.5213	242.6546	248.0315
	5.0	88.2900	118.6742	165.4517	201.3055	227.7422	232.4898
SCSS	0.1	196.1879	263.1995	382.1383	555.4219	568.4830	623.4845
	0.2	155.1037	205.9736	294.8680	422.1441	433.0060	472.8239
	0.5	124.0104	162.1799	226.9125	316.2627	325.7018	352.7728
	1.0	111.6732	144.6280	199.1561	271.9382	280.9621	302.3248
	2.0	104.9183	134.9734	183.6997	246.7925	255.6670	273.6209
	5.0	100.6169	128.8152	173.7579	230.3777	239.2047	254.8399
SSSF	0.1	66.1606	143.6928	226.7119	270.5245	313.9389	453.6732
	0.2	59.7244	118.6314	185.1117	213.1828	247.9254	350.1790
	0.5	55.3189	100.5444	154.3466	169.6662	197.7382	269.5104
	1.0	53.6878	93.6858	142.2362	152.3853	177.6664	236.4694
	2.0	52.8237	90.0350	135.5655	142.9423	166.5952	217.6501
	5.0	52.2850	87.7582	131.2764	136.9515	159.5016	205.0581
SCSC	0.1	258.9497	314.8591	421.6898	598.2579	670.5301	730.9672
	0.2	199.5450	242.6829	323.2007	454.5346	505.8012	550.9948
	0.5	153.1617	186.3754	245.8248	340.3062	373.5551	406.3759
	1.0	134.1289	163.3112	213.8793	292.4687	317.4225	344.9199
	2.0	123.4774	150.4315	195.9472	265.3274	285.2155	309.6282
	5.0	116.5898	142.1234	184.3400	247.6126	263.9894	286.3541
SCSF	0.1	81.5024	153.6340	277.5241	286.5253	367.0596	470.6596
	0.2	70.7141	125.8310	218.3040	228.1729	286.5358	361.2035
	0.5	62.8196	105.4833	173.2282	183.8452	224.3794	275.3381
	1.0	59.6703	97.6302	155.2563	165.9181	199.0632	239.9440
	2.0	57.9107	93.3909	145.4023	155.8512	184.8980	220.0975
	5.0	56.7674	90.7160	139.1323	149.2724	175.7101	207.2619
SFSF	0.1	43.7108	109.2055	121.2945	194.7775	247.5588	313.1839
	0.2	40.4679	97.9095	101.2535	159.3344	195.8936	254.0387
	0.5	38.3764	87.0079	90.0349	133.3982	156.9023	204.9344
	1.0	37.6471	81.6897	86.9762	123.3794	141.5167	182.8535
	2.0	37.2748	78.8875	85.2661	117.9573	133.1472	170.6786
	5.0	37.0486	77.1528	84.1445	114.5238	127.8558	162.8920

Source: Reproduced from Pradhan, K.K., Chakraverty, S., Free vibration of FG Lévy plate resting on elastic foundations. Engineering and Computational Mechanics 2015;1500014 (26 pages).

Table 8.14 Effect of ρ_r on the first six non-dimensional frequencies of functionally graded Lévy plates; $K_W = 0$, $K_S = 0$, $k = 1$, $\mu = 2.0$, and $E_r = 3.0$

BCs	ρ_r	λ_1	λ_2	λ_3	λ_4	λ_5	λ_6
SSSS	0.1	17.1808	27.4896	44.6778	58.4160	68.7293	70.9060
	0.2	23.2629	37.2211	60.4941	79.0956	93.0600	96.0072
	0.5	32.8987	52.6386	85.5515	111.8581	131.6067	135.7747
	1.0	40.2925	64.4688	104.7788	136.9976	161.1846	166.2894
	2.0	46.5258	74.4422	120.9881	158.1912	186.1199	192.0145
	5.0	52.0174	83.2289	135.2688	176.8632	208.0884	214.6787
SCSS	0.1	24.1366	32.9305	48.8563	72.5528	74.0394	81.6901
	0.2	32.6811	44.5881	66.1518	98.2369	100.2498	110.6089
	0.5	46.2180	63.0571	93.5527	138.9279	141.7746	156.4246
	1.0	56.6053	77.2289	114.5782	170.1513	173.6377	191.5802
	2.0	65.3621	89.1763	132.3036	196.4738	200.4996	221.2177
	5.0	73.0771	99.7021	147.9199	219.6644	224.1653	247.3290
SSSF	0.1	5.6174	16.2726	26.2104	33.4670	38.6668	57.9587
	0.2	7.6060	22.0332	35.4891	45.3145	52.3551	78.4764
	0.5	10.7565	31.1597	50.1891	64.0844	74.0413	110.9824
	1.0	13.1740	38.1627	61.4689	78.4870	90.6817	135.9252
	2.0	15.2120	44.0665	70.9781	90.6290	104.7102	156.9529
	5.0	17.0075	49.2678	79.3560	101.3263	117.0695	175.4787
SCSC	0.1	33.1661	40.3177	54.4417	78.1882	88.4796	96.5470
	0.2	44.9072	54.5904	73.7144	105.8673	119.8020	130.7252
	0.5	63.5083	77.2025	104.2479	149.7189	169.4255	184.8734
	1.0	77.7815	94.5534	127.6770	183.3675	207.5031	226.4227
	2.0	89.8144	109.1808	147.4287	211.7345	239.6039	261.4505
	5.0	100.4155	122.0679	164.8303	236.7264	267.8853	292.3105
SCSF	0.1	7.9433	17.6692	34.4289	34.7373	46.0568	60.6475
	0.2	10.7553	23.9242	46.6170	47.0345	62.3612	82.1170
	0.5	15.2103	33.8339	65.9263	66.5168	88.1920	116.1310
	1.0	18.6287	41.4379	80.7429	81.4661	108.0127	142.2309
	2.0	21.5106	47.8484	93.2339	94.0689	124.7224	164.2341
	5.0	24.0496	53.4962	104.2387	105.1723	139.4438	183.6193
SFSF	0.1	3.3118	9.5820	13.4149	22.4732	30.4200	36.7277
	0.2	4.4842	12.9741	18.1639	30.4289	41.1889	49.7296
	0.5	6.3416	18.3482	25.6876	43.0330	58.2499	70.3282
	1.0	7.7669	22.4718	31.4608	52.7044	71.3413	86.1341
	2.0	8.9684	25.9482	36.3278	60.8578	82.3778	99.4591
	5.0	10.0270	29.0110	40.6157	68.0411	92.1012	111.1987

Source: Reproduced from Pradhan, K.K., Chakraverty, S., Free vibration of FG Lévy plate resting on elastic foundations. Engineering and Computational Mechanics 2015;1500014 (26 pages).

Table 8.15 Effect of ρ_r on the first six non-dimensional frequencies of functionally graded Lévy plates; $K_W = 100$, $K_S = 0$, $k = 1$, $\mu = 2.0$, and $E_r = 3.0$

BCs	ρ_r	λ_1	λ_2	λ_3	λ_4	λ_5	λ_6
SSSS	0.1	17.7020	27.8183	44.8808	58.5714	68.8615	71.0341
	0.2	23.9686	37.6662	60.7689	79.3060	93.2389	96.1807
	0.5	33.8968	53.2681	85.9403	112.1557	131.8597	136.0200
	1.0	41.5149	65.2398	105.2549	137.3621	161.4945	166.5898
	2.0	47.9372	75.3324	121.5379	158.6121	186.4778	192.3614
	5.0	53.5955	84.2242	135.8835	177.3337	208.4885	215.0665
SCSS	0.1	24.5103	33.2054	49.0421	72.6779	74.1621	81.8013
	0.2	33.1871	44.9604	66.4033	98.4064	100.4159	110.7595
	0.5	46.9337	63.5836	93.9084	139.1677	142.0095	156.6375
	1.0	57.4818	77.8736	115.0138	170.4449	173.9254	191.8410
	2.0	66.3743	89.9207	132.8065	196.8128	200.8318	221.5189
	5.0	74.2087	100.5344	148.4822	220.0434	224.5368	247.6657
SSSF	0.1	7.0525	16.8220	26.5550	33.7375	38.9012	58.1153
	0.2	9.5491	22.7771	35.9556	45.6808	52.6725	78.6885
	0.5	13.5044	32.2117	50.8489	64.6024	74.4901	111.2824
	1.0	16.5395	39.4511	62.2770	79.1215	91.2314	136.2925
	2.0	19.0981	45.5542	71.9113	91.3616	105.3449	157.3771
	5.0	21.3524	50.9312	80.3992	102.1454	117.7792	175.9529
SCSC	0.1	33.4391	40.5425	54.6084	78.3043	88.5823	96.6411
	0.2	45.2768	54.8949	73.9401	106.0246	119.9410	130.8527
	0.5	64.0311	77.6331	104.5671	149.9414	169.6222	185.0536
	1.0	78.4217	95.0807	128.0680	183.6400	207.7439	226.6435
	2.0	90.5536	109.7897	147.8802	212.0492	239.8820	261.7053
	5.0	101.2420	122.7486	165.3351	237.0782	268.1962	292.5955
SCSF	0.1	9.0154	18.1764	34.6920	34.9980	46.2537	60.7972
	0.2	12.2070	24.6110	46.9731	47.3875	62.6279	82.3197
	0.5	17.2633	34.8052	66.4300	67.0160	88.5692	116.4177
	1.0	21.1431	42.6275	81.3598	82.0776	108.4747	142.5820
	2.0	24.4139	49.2220	93.9462	94.7750	125.2557	164.6395
	5.0	27.2956	55.0319	105.0351	105.9617	140.0402	184.0726
SFSF	0.1	5.3991	10.4879	14.0763	22.8742	30.7174	36.9744
	0.2	7.3104	14.2007	19.0594	30.9718	41.5916	50.0636
	0.5	10.3384	20.0829	26.9540	43.8007	58.8194	70.8006
	1.0	12.6619	24.5964	33.0118	53.6447	72.0387	86.7127
	2.0	14.6207	28.4015	38.1188	61.9436	83.1831	100.1272
	5.0	16.3465	31.7538	42.6181	69.2550	93.0016	111.9456

Source: Reproduced from Pradhan, K.K., Chakraverty, S., Free vibration of FG Lévy plate resting on elastic foundations. Engineering and Computational Mechanics 2015;1500014 (26 pages).

Table 8.16 Effect of ρ_r on the first six non-dimensional frequencies of functionally graded Lévy plates; $K_W = 0$, $K_S = 100$, $k = 1$, $\mu = 2.0$, and $E_r = 3.0$

BCs	ρ_r	λ_1	λ_2	λ_3	λ_4	λ_5	λ_6
SSSS	0.1	34.5314	46.8109	65.7955	80.3931	91.1743	93.1362
	0.2	46.7557	63.3823	89.0875	108.8527	123.4506	126.1071
	0.5	66.1225	89.6361	125.9887	153.9410	174.5855	178.3423
	1.0	80.9833	109.7813	154.3040	188.5384	213.8227	218.4239
	2.0	93.5114	126.7645	178.1750	217.7054	246.9012	252.2141
	5.0	104.5489	141.7270	199.2057	243.4021	276.0439	281.9840
SCSS	0.1	39.6910	51.0436	69.2511	92.4711	95.9166	102.4217
	0.2	53.7419	69.1134	93.7664	125.2064	129.8717	138.6796
	0.5	76.0025	97.7411	132.6057	177.0686	183.6663	196.1226
	1.0	93.0837	119.7079	162.4081	216.8639	224.9443	240.2001
	2.0	107.4837	138.2267	187.5328	250.4129	259.7434	277.3592
	5.0	120.1705	154.5422	209.6680	279.9701	290.4019	310.0970
SSSF	0.1	19.9963	34.2930	51.6779	54.1901	63.1881	81.9230
	0.2	27.0752	46.4330	69.9722	73.3738	85.5570	110.9243
	0.5	38.2901	65.6661	98.9557	103.7662	120.9959	156.8706
	1.0	46.8956	80.4243	121.1955	127.0871	148.1891	192.1265
	2.0	54.1503	92.8659	139.9445	146.7476	171.1141	221.8485
	5.0	60.5419	103.8273	156.4627	164.0688	191.3114	248.0342
SCSC	0.1	46.3991	56.6222	73.6738	99.4156	106.4248	115.4996
	0.2	62.8247	76.6669	99.7548	134.6094	144.0998	156.3872
	0.5	88.8476	108.4233	141.0746	190.3665	203.7879	221.1648
	1.0	108.8156	132.7909	172.7804	233.1504	249.5882	270.8705
	2.0	125.6494	153.3337	199.5096	269.2188	288.1997	312.7743
	5.0	140.4803	171.4323	223.0585	300.9958	322.2170	349.6923
SCSF	0.1	21.8867	35.5243	55.0915	59.1123	69.8838	82.8255
	0.2	29.6348	48.1001	74.5942	80.0385	94.6231	112.1462
	0.5	41.9099	68.0238	105.4922	113.1915	133.8173	158.5987
	1.0	51.3289	83.3118	129.2010	138.6307	163.8921	194.2429
	2.0	59.2695	96.2002	149.1885	160.0769	189.2463	224.2924
	5.0	66.2653	107.5551	166.7978	178.9714	211.5838	250.7665
SFSF	0.1	13.8174	30.0327	32.6745	44.9782	50.5142	64.6319
	0.2	18.7089	40.6644	44.2414	60.9007	68.3966	87.5120
	0.5	26.4583	57.5082	62.5668	86.1266	96.7274	123.7607
	1.0	32.4047	70.4328	76.6284	105.4831	118.4663	151.5753
	2.0	37.4178	81.3288	88.4828	121.8014	136.7931	175.0240
	5.0	41.8343	90.9284	98.9268	136.1781	152.9394	195.6828

Source: Reproduced from Pradhan, K.K., Chakraverty, S., Free vibration of FG Lévy plate resting on elastic foundations. Engineering and Computational Mechanics 2015;1500014 (26 pages).

Table 8.17 Effect of ρ_r on the first six non-dimensional frequencies of functionally graded Lévy plates; $K_W = 100$, $K_S = 100$, $k = 1$, $\mu = 2.0$, and $E_r = 3.0$

BCs	ρ_r	λ_1	λ_2	λ_3	λ_4	λ_5	λ_6
SSSS	0.1	34.7936	47.0047	65.9335	80.5061	91.2739	93.2338
	0.2	47.1108	63.6447	89.2744	109.0057	123.5855	126.2392
	0.5	66.6248	90.0072	126.2530	154.1574	174.7763	178.5291
	1.0	81.5983	110.2358	154.6277	188.8035	214.0564	218.6526
	2.0	94.2216	127.2893	178.5487	218.0115	247.1710	252.4783
	5.0	105.3430	142.3138	199.6236	243.7442	276.3456	282.2794
SCSS	0.1	39.9194	51.2214	69.3822	92.5693	96.0113	102.5104
	0.2	54.0511	69.3541	93.9440	125.3395	129.9999	138.7997
	0.5	76.4398	98.0815	132.8568	177.2568	183.8477	196.2925
	1.0	93.6193	120.1248	162.7157	217.0943	225.1665	240.4082
	2.0	108.1022	138.7082	187.8879	250.6790	259.9999	277.5995
	5.0	120.8620	155.0805	210.0651	280.2676	290.6887	310.3656
SSSF	0.1	20.4459	34.5571	51.8535	54.3576	63.3318	82.0339
	0.2	27.6839	46.7905	70.2100	73.6006	85.7516	111.0744
	0.5	39.1509	66.1718	99.2920	104.0869	121.2711	157.0829
	1.0	47.9499	81.0436	121.6073	127.4800	148.5262	192.3865
	2.0	55.3678	93.5811	140.4201	147.2012	171.5032	222.1488
	5.0	61.9031	104.6268	156.9944	164.5759	191.7465	248.3699
SCSC	0.1	46.5946	56.7826	73.7971	99.5071	106.5102	115.5782
	0.2	63.0894	76.8840	99.9217	134.7332	144.2154	156.4937
	0.5	89.2220	108.7303	141.3107	190.5415	203.9514	221.3155
	1.0	109.2741	133.1669	173.0695	233.3647	249.7885	271.0550
	2.0	126.1789	153.7679	199.8434	269.4664	288.4309	312.9874
	5.0	141.0723	171.9178	223.4318	301.2726	322.4755	349.9305
SCSF	0.1	22.2982	35.7793	55.2563	59.2659	70.0138	82.9351
	0.2	30.1919	48.4454	74.8173	80.2464	94.7991	112.2947
	0.5	42.6978	68.5121	105.8077	113.4856	134.0662	158.8087
	1.0	52.2939	83.9098	129.5874	138.9909	164.1969	194.5001
	2.0	60.3839	96.8907	149.6347	160.4928	189.5982	224.5894
	5.0	67.5112	108.3271	167.2967	179.4365	211.9773	251.0986
SFSF	0.1	14.4604	30.3338	32.9515	45.1798	50.6939	64.7724
	0.2	19.5795	41.0722	44.6166	61.1738	68.6398	87.7023
	0.5	27.6895	58.0849	63.0973	86.5128	97.0714	124.0297
	1.0	33.9126	71.1392	77.2781	105.9561	118.8876	151.9048
	2.0	39.1589	82.1444	89.2331	122.3475	137.2796	175.4045
	5.0	43.7810	91.8403	99.7656	136.7887	153.4833	196.1082

Source: Reproduced from Pradhan, K.K., Chakraverty, S., Free vibration of FG Lévy plate resting on elastic foundations. Engineering and Computational Mechanics 2015;1500014 (26 pages).

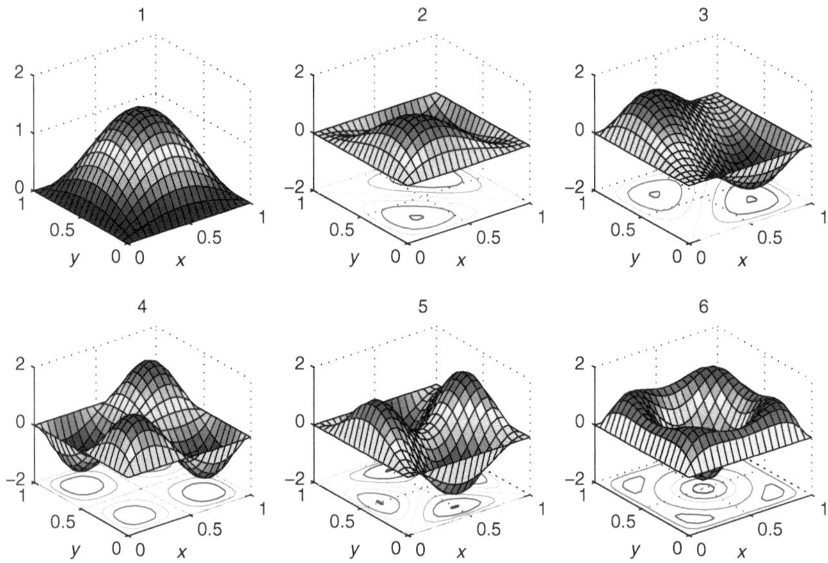

Figure 8.2 The first six 3D mode shapes of SSSS Al/Al_2O_3 Lévy square plates with $k = 1$, $K_W = 100$, and $K_S = 100$. (*Reproduced from Pradhan, K.K., Chakraverty, S., Free vibration of FG Lévy plate resting on elastic foundations. Engineering and Computational Mechanics 2015;1500014 (26 pages).*).

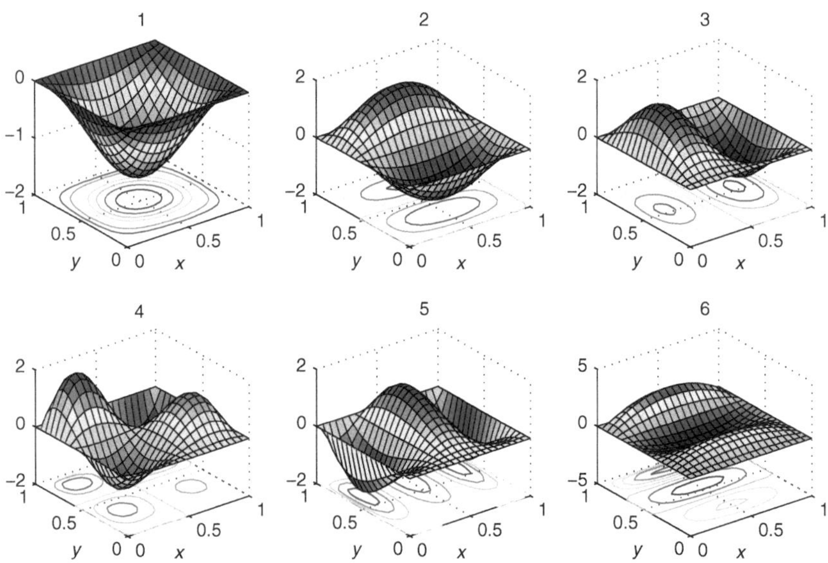

Figure 8.3 The first six 3D mode shapes of SCSS Al/Al_2O_3 Lévy square plates with $k = 1$, $K_W = 100$, and $K_S = 100$. (*Reproduced from Pradhan, K.K., Chakraverty, S., Free vibration of FG Lévy plate resting on elastic foundations. Engineering and Computational Mechanics 2015;1500014 (26 pages).*).

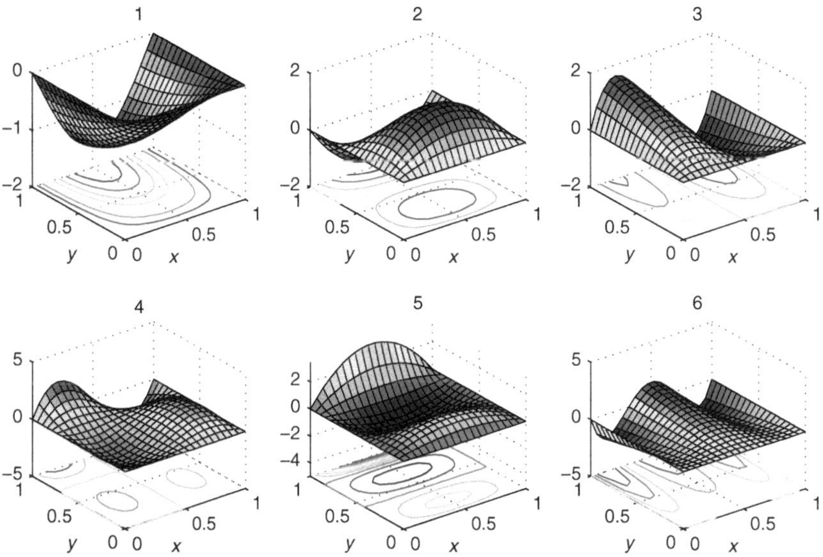

Figure 8.4 The first six 3D mode shapes of SSSF Al/Al_2O_3 Lévy square plates with $k = 1$, $K_W = 100$, and $K_S = 100$. *(Reproduced from Pradhan, K.K., Chakraverty, S., Free vibration of FG Lévy plate resting on elastic foundations. Engineering and Computational Mechanics 2015;1500014 (26 pages).)*.

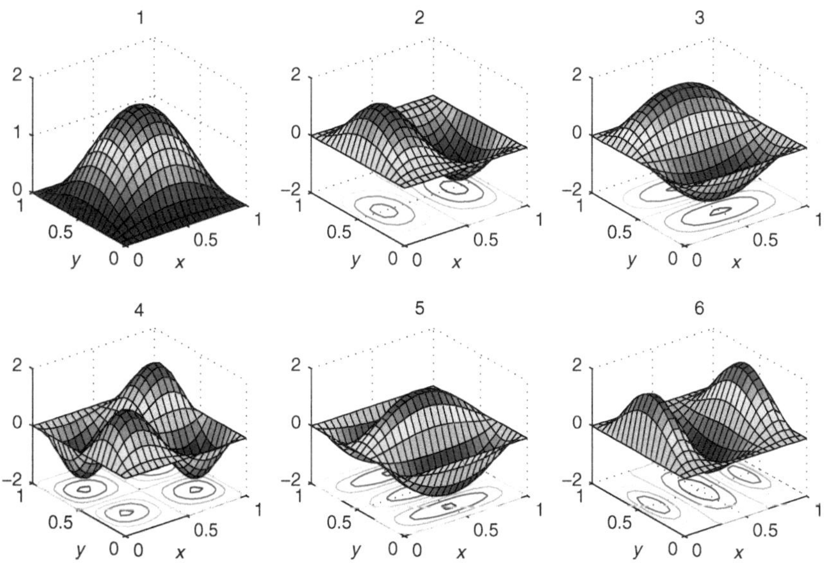

Figure 8.5 The first six 3D mode shapes of SCSC Al/Al_2O_3 Lévy square plates with $k = 1$, $K_W = 100$, and $K_S = 100$. *(Reproduced from Pradhan, K.K., Chakraverty, S., Free vibration of FG Lévy plate resting on elastic foundations. Engineering and Computational Mechanics 2015;1500014 (26 pages).)*.

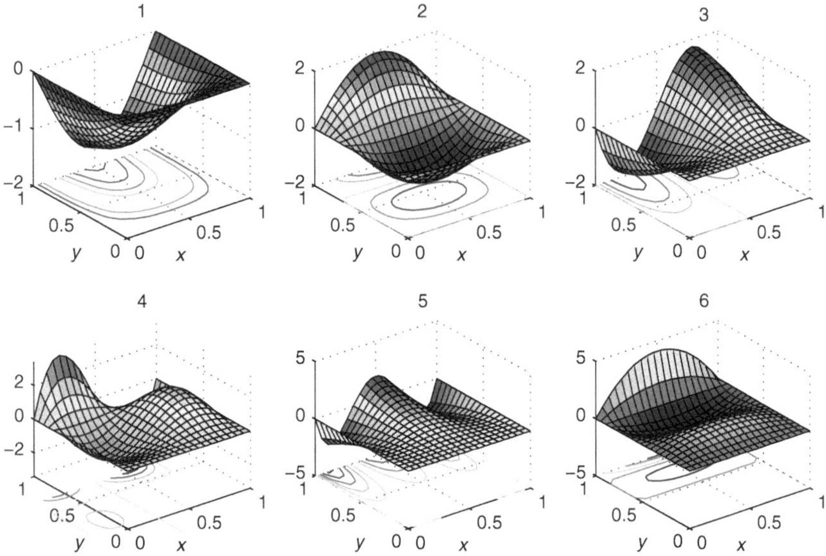

Figure 8.6 The first six 3D mode shapes of SCSF Al/Al_2O_3 Lévy square plates with $k = 1$, $K_W = 100$, and $K_S = 100$. (*Reproduced from Pradhan, K.K., Chakraverty, S., Free vibration of FG Lévy plate resting on elastic foundations. Engineering and Computational Mechanics 2015;1500014 (26 pages).*).

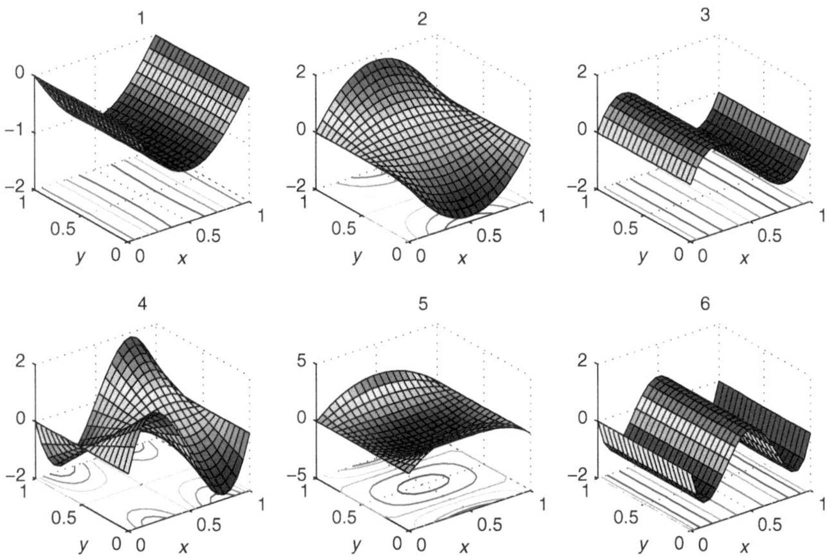

Figure 8.7 The first six 3D mode shapes of SFSF Al/Al_2O_3 Lévy square plates with $k = 1$, $K_W = 100$, and $K_S = 100$. (*Reproduced from Pradhan, K.K., Chakraverty, S., Free vibration of FG Lévy plate resting on elastic foundations. Engineering and Computational Mechanics 2015;1500014 (26 pages).*).

along the thickness direction. Accordingly, the results for exponential FG plates under the consideration of thermal conditions are incorporated after checking the convergence of frequencies and comparison studies in special cases. The effect of temperature gradients on free vibration eigenfrequencies has also been estimated, as done in Chakraverty and Pradhan (2014a).

8.2.1 Temperature Field Across the Thickness

The main objective of this investigation is to study the free vibration of FG plates (exponential gradation behavior is supposed for the volume fractions) in thermal environments. As such, let us first discuss the two different cases of temperature rise and its impact on mechanical properties. The temperature variation is assumed to occur in thickness direction only, and one-dimensional temperature is considered to be constant in the plane of the plate. As mentioned in Li et al. (2009a) and Shi and Dong (2012), two cases of temperature gradation are assumed across the thickness.

Uniform Temperature Rise

The temperature field across the thickness in the case of uniform temperature rise can be expressed as

$$T = T_0 + \Delta T \tag{8.7}$$

Here, T_0 and ΔT denote the respective initial uniform temperature of the plate and temperature change in Eq. (8.7).

Linear Temperature Rise

Let us assume that T_c and T_m are the temperatures at the top and bottom face of the FG plate. The temperature variation is now a linear function involving the thickness coordinate (z). As such, the temperature field under linear temperature rise can be written as

$$T(z) = T_m + \Delta T\left(\frac{z}{h} + \frac{1}{2}\right) \tag{8.8}$$

In Eq. (8.8), $\Delta T = T_c - T_m$ is the temperature gradient and $T_m = 300$ K represents the initial temperature. The material properties that is Young's modulus (E), Poisson's ratio (ν), and thermal expansion coefficient (α) are nonlinear functions of temperature as stated in Reddy and Chin (1998), Shen (2002), Yang and Shen (2002), Kim (2005),

Kitipornchai et al. (2006), Li et al. (2009a), Malekzadeh and Beni (2010), and Shi and Dong (2012):

$$\mathcal{P}(T) = \mathcal{P}_0^T \left(\mathcal{P}_{-1}^T T^{-1} + 1 + \mathcal{P}_1^T T + \mathcal{P}_2^T T^2 + \mathcal{P}_3^T T^3\right) \tag{8.9}$$

where $T = T_0 + \Delta T(z)$ denotes the environmental temperature and T_0 is the free stress or room temperature; $\mathcal{P}_{-1}^T$, $\mathcal{P}_0^T$, $\mathcal{P}_1^T$, $\mathcal{P}_2^T$, and $\mathcal{P}_3^T$ are the constant coefficients in the cubic expression of the respective temperature-dependent material property.

It can easily be noticed in Reddy and Chin (1998), Shen (2002), Yang and Shen (2002), Kim (2005), Kitipornchai et al. (2006), Li et al. (2009a), Malekzadeh and Beni (2010), and Shi and Dong (2012) that the effect of thermal environment in FG plates assumes the power-law variation of material properties only. The present study may be the first where we assume the corresponding exponential gradation of material properties in thermal environments as

$$\begin{aligned} E(z,T) &= E_c(T)\mathrm{e}^{-\delta\left(1-\frac{2z}{h}\right)};\, \delta = \frac{1}{2}\ln\left(\frac{E_c(T)}{E_m(T)}\right) \\ \alpha(z,T) &= \alpha_c(T)\mathrm{e}^{-\delta\left(1-\frac{2z}{h}\right)};\, \delta = \frac{1}{2}\ln\left(\frac{\alpha_c(T)}{\alpha_m(T)}\right) \\ \nu(z,T) &= \nu_c(T)\mathrm{e}^{-\delta\left(1-\frac{2z}{h}\right)};\, \delta = \frac{1}{2}\ln\left(\frac{\nu_c(T)}{\nu_m(T)}\right) \end{aligned} \tag{8.10}$$

One can observe in Table 8.18 that only mass density (ρ) includes non-zero value only in the coefficient viz. $\mathcal{P}_0^T$ (free stress state). Hence mass density will vary exponentially without involving the effect of temperature in its expression.

8.2.2 Numerical Modeling

State Under Temperature Gradient

Initially, let us assume the free stress temperature be $T_0 = 300$ K. Now, the plate is acted upon by a thermal gradient, ΔT (either uniform or linear temperature rise) (Kim, 2005). As a result, the constitutive relations can be written as

$$d_{ij} = \frac{\partial u_x}{\partial i}\frac{\partial u_x}{\partial j} + \frac{\partial u_y}{\partial i}\frac{\partial u_y}{\partial j} + \frac{\partial u_z}{\partial i}\frac{\partial u_z}{\partial j} \quad (i,j = x,y) \tag{8.11}$$

Table 8.18 Temperature-dependent coefficients of Young's modulus (E), Poisson's ratio (ν), thermal expansion coefficient (α) and mass density (ρ) of Si_3N_4 and SUS304 as mentioned in (Kim, 2005)

Properties (Unit)	Material	$\mathcal{P}_{-1}^T$	$\mathcal{P}_0^T$	$\mathcal{P}_1^T$	$\mathcal{P}_2^T$	$\mathcal{P}_3^T$	$\mathcal{P}$ (at $T = 300$ K)
E (GPa)	SUS304	0	201.04	3.079×10^{-4}	-6.534×10^{-7}	0	207.7877
	Si_3N_4	0	348.43	-3.070×10^{-4}	2.160×10^{-7}	-8.946×10^{-11}	322.2715
ν (no unit)	SUS304	0	0.3262	-2.002×10^{-4}	3.797×10^{-7}	0	0.3178
	Si_3N_4	0	0.2400	0	0	0	0.2400
α (1/K)	SUS304	0	12.330×10^{-6}	8.086×10^{-6}	0	0	1.5321×10^{-5}
	Si_3N_4	0	5.8723×10^{-6}	9.095×10^{-6}	0	0	7.4746×10^{-6}
ρ (kg/m^3)	SUS304	0	8166	0	0	0	8166
	Si_3N_4	0	2370	0	0	0	2370

Source: Reproduced from Chakraverty, S., Pradhan, K.K., Free vibration of exponential functionally graded rectangular plates in thermal environment with general boundary conditions. Aerospace Science and Technology 2014;36:132–156.

In Eq. (8.11), d_{ij} are the components of strains due to temperature gradation. Consequently, corresponding thermal stresses can be defined in matrix form as (Kim, 2005; Kitipornchai et al., 2006)

$$\begin{Bmatrix} \sigma_{xx}^T \\ \sigma_{yy}^T \\ \tau_{xy}^T \end{Bmatrix} = -\begin{pmatrix} Q_{11} & Q_{12} & 0 \\ Q_{21} & Q_{22} & 0 \\ 0 & 0 & Q_{66} \end{pmatrix} \begin{pmatrix} 1 & 0 \\ 0 & 1 \\ 0 & 0 \end{pmatrix} \begin{Bmatrix} \alpha(z,T) \\ \alpha(z,T) \end{Bmatrix} \Delta T(z) \tag{8.12}$$

In Table 8.18, the coefficients for Poisson's ratio in cubic expansion vary with a certain temperature. So, the reduced stiffness coefficients in Eq. (8.12) will be

$$Q_{11} = Q_{22} = \frac{E(z,T)}{1-\nu^2(z,T)}, \quad Q_{12} = Q_{21} = \frac{\nu(z,T)E(z,T)}{1-\nu^2(z,T)},$$
$$Q_{66} = \frac{E(z,T)}{2[1+\nu(z,T)]}$$

Energy Expressions in Thermal Environment

Let $\mathbf{U}_T$ be the strain energy from the thermal stresses (as stated in Eq. (8.12)) due to temperature rise. Then $\mathbf{U}_T$ can be given as

$$\mathbf{U}_T = \frac{1}{2}\int_V \left(\sigma_{xx}^T d_{xx} + \sigma_{yy}^T d_{yy} + \tau_{xy}^T d_{xy}\right) \mathrm{d}V \tag{8.13}$$

where

$$\sigma_{xx}^T = -(Q_{11} + Q_{12})\,\alpha(z,T)\Delta T(z),$$
$$\sigma_{yy}^T = -(Q_{12} + Q_{22})\,\alpha(z,T)\Delta T(z),$$
$$\tau_{xy}^T = 0$$

and

$$d_{xx} = z^2\left\{\left(\frac{\partial^2 w}{\partial x^2}\right)^2 + \left(\frac{\partial^2 w}{\partial x \partial y}\right)^2\right\} + \left(\frac{\partial w}{\partial x}\right)^2$$
$$d_{yy} = z^2\left\{\left(\frac{\partial^2 w}{\partial x \partial y}\right)^2 + \left(\frac{\partial^2 w}{\partial y^2}\right)^2\right\} + \left(\frac{\partial w}{\partial y}\right)^2$$
$$d_{xy} = z^2\left(\frac{\partial^2 w}{\partial x^2} + \frac{\partial^2 w}{\partial y^2}\right)\frac{\partial^2 w}{\partial x \partial y} + \frac{\partial w}{\partial x}\frac{\partial w}{\partial y}$$

Substituting these terms in Eq. (8.13), leads to

$$\mathbf{U}_T = -\frac{1}{2}\int_{\Omega}\left[\left(D_{11}^T + D_{12}^T\right)\left\{\left(\frac{\partial^2 w}{\partial x^2}\right)^2 + 2\left(\frac{\partial^2 w}{\partial x \partial y}\right)^2 + \left(\frac{\partial^2 w}{\partial y^2}\right)^2\right\} + \left(A_{11}^T + A_{12}^T\right)\left\{\left(\frac{\partial w}{\partial x}\right)^2 + \left(\frac{\partial w}{\partial y}\right)^2\right\}\right] \mathrm{d}x\,\mathrm{d}y \tag{8.14}$$

Here, thermal stiffness coefficients in Eq. (8.14) are defined as

$$\left(D_{11}^T, D_{12}^T\right) = \int_{-h/2}^{h/2} (Q_{11}, Q_{12})\, z^2 \alpha \Delta T\, \mathrm{d}z$$

$$\left(A_{11}^T, A_{12}^T\right) = \int_{-h/2}^{h/2} (Q_{11}, Q_{12})\, \alpha \Delta T\, \mathrm{d}z$$

Using harmonic type deflection, $w(x, y, t) = W(x, y)\cos\omega t$, Eq. (8.14) yields maximum thermal strain energy, $(\mathbf{U}_T)_{\max}$ as

$$(\mathbf{U}_T)_{\max} = -\frac{1}{2}\int_{\Omega}\left[\left(D_{11}^T + D_{12}^T\right)\left\{\left(\frac{\partial^2 W}{\partial x^2}\right)^2 + 2\left(\frac{\partial^2 W}{\partial x \partial y}\right)^2 + \left(\frac{\partial^2 W}{\partial y^2}\right)^2\right\} + \left(A_{11}^T + A_{12}^T\right)\left\{\left(\frac{\partial W}{\partial x}\right)^2 + \left(\frac{\partial W}{\partial y}\right)^2\right\}\right] \mathrm{d}x\,\mathrm{d}y \tag{8.15}$$

Due to temperature rise or thermal environment, the effective strain energy ($\mathbf{U}_{\text{eff}}$) will be obtained using expressions involved in Eq. (3.19) and (8.15) as $\mathbf{U}_{\text{eff}} = \mathbf{U}_{\max} + (\mathbf{U}_T)_{\max}$, where $\mathbf{U}_{\max} = \frac{1}{2}\int_{\Omega}\left[D_{11}\left\{\left(\frac{\partial^2 W}{\partial x^2}\right)^2 + \left(\frac{\partial^2 W}{\partial y^2}\right)^2\right\} + 2D_{12}\frac{\partial^2 W}{\partial x^2}\frac{\partial^2 W}{\partial y^2} + 4D_{66}\left(\frac{\partial^2 W}{\partial x \partial y}\right)^2\right] \mathrm{d}x\,\mathrm{d}y$. In the case of FG plates in a thermal environment, the Rayleigh-Ritz method is used by equating $\mathbf{U}_{\text{eff}}$ and maximum kinetic energy ($\mathbf{T}_{\max}$) to get the Rayleigh quotient (ω^2). Differentiate this quotient with respect to the unknown constants (c_i; $i = 1, 2, 3, \ldots, n$), and the generalized eigenvalue problem for exponential FG plate in a thermal environment of the form Eq. (3.26) can easily be obtained.

8.2.3 Results and Discussions

The target of this work is to study the free vibration of exponential FG plate in thermal environments. To validate our results in thermal condition, we could not find any paper related to exponential FG plate with thermal environments. But one can have papers concerned with FG plates having power-law variations of temperature-dependent material properties. So, we compute results for power-law-based FG plates in thermal environments and as such, a convergence study is performed in Table 8.19 for the first eight natural frequencies of CCCC FG plates with power-law variation of material properties. The convergence of free vibration frequencies of clamped exponential FG plates under uniform and linear temperature environments is tabulated in Table 8.20. The material properties under initial stress free temperature ($T = 300\,\mathrm{K}$) are considered (as stated in Table 8.18). The power-law exponents (k) are taken as 2.0 and 10.0 in both uniform and linear temperature conditions; other physical parameters are $a = 0.2\,\mathrm{m}$, $h/b = 0.1$, and temperature gradient (ΔT) $= 300\,\mathrm{K}$. The first eight non-dimensional frequencies are evaluated with two different aspect ratios, $a/b = 0.5$ and 1.0. Frequencies associated with uniform rise of temperature condition for power-law based FG plate are validated with Yang and Shen (2002) and Li et al. (2009a).

In validation, one may observe discrepancies in our results. This is due to the fact that transverse shear deformation effects are ignored in classical plate theory, based on which the present investigation is performed. The Diff. % used in the table is the percentage differences of the computed results (CPT) and the results taken from the available literature. It can be viewed that the Diff. % is significant for $a/b = 0.5$, whereas this is insignificant for $a/b = 1$ except at higher modes. Regardless of the thermal condition and variation form assumed, one may say that the number of polynomials (n) involved in deflection function (W) plays a key role for the convergence of frequencies at each mode. It can also easily be noticed that frequencies associated with linear temperature rise is higher than that for uniform rise of temperature.

The effects of aspect ratios on the first eight non-dimensional frequencies of exponential FG (SUS304/Si_3N_4) plate under two different thermal environments are incorporated in Tables 8.21–8.24. Here also, the material properties of FG constituents are taken as mentioned in Table 8.18. Only clamped and simply supported edge supports are considered here. Tables 8.21 and 8.22 are meant for CCCC FG plates subjected to uniform and linear temperature rise, respectively, whereas similar assumptions are considered in Tables 8.23 and 8.24 for FG plate with SSSS edge support. Initial free stress or room temperature (T) is taken as 300 K and temperature gradient (ΔT) are assumed to be 0, 300, and 500 for each aspect ratio (a/b). Physical parameters involved in these computations are: $a/b = 0.5$ and 1.0, h/b (thickness-to-breadth ratio) $= 0.1,\ 0.2,$ and 0.25. Looking into the

Table 8.19 Convergence and comparison of the first eight natural frequencies of CCCC FG ($SUS304/Si_3N_4$) plates with different thermal conditions ($a = 0.2$ m, $h/b = 0.1$, $T = 300$ K, $\Delta T = 300$ K)

a/b (thermal state)	*k*	Source	λ_1	λ_2	λ_3	λ_4	λ_5	λ_6	λ_7	λ_8
0.5 (uniform rise)	2.0	8 × 8	9.9116	12.8535	18.4206	25.8682	29.0255	52.4124	55.0152	90.4206
		10 × 10	9.9116	12.8418	18.4206	25.8243	27.2368	29.0255	33.8882	52.4124
		13 × 13	9.9106	12.8418	18.4095	25.8243	27.2368	28.7053	33.8882	49.8405
		15 × 15	9.9101	12.8418	18.0820	25.8243	27.2368	28.6941	33.8882	40.0102
		Yang and Shen (2002)	9.2196	11.6913	15.2957	20.4667	21.2323	21.4468	22.4853	25.4461
		Diff. %	7.5057	9.8407	18.2162	23.5387	28.2800	33.7919	50.7127	57.2351
		Li et al. (2009a)	9.2111	11.5890	15.5999	19.9043	20.0234	20.7922	21.9073	25.0100
		Diff. %	7.6049	10.8103	15.9109	29.7423	36.0249	38.0042	54.6891	59.9768
	10.0	8 × 8	9.9029	12.8433	18.4066	25.8473	29.0036	52.3713	54.9737	90.3504
		10 × 10	9.9029	12.8316	18.4066	25.8034	27.2164	29.0036	33.8643	52.3713
		13 × 13	9.9020	12.8316	18.3955	25.8034	27.2164	28.6837	33.8643	49.8012
		15 × 15	9.9016	12.8316	18.0683	25.8034	27.2164	28.6725	33.8643	39.9803
		Yang and Shen (2002)	7.9839	10.1219	13.3088	17.6295	18.3727	18.9066	19.3778	21.9914
		Li et al. (2009a)	7.8170	9.8332	13.2410	16.8733	16.9943	17.6538	18.5960	21.2369
1.0 (uniform rise)	2.0	8 × 8	3.6322	7.4211	7.4848	10.9601	13.8504	13.9879	17.0379	23.3609
		10 × 10	3.6322	7.4092	7.4092	10.9601	13.8504	13.9879	17.0374	17.0374
		13 × 13	3.6312	7.4092	7.4092	10.9437	13.3312	13.8915	17.0374	17.0374
		15 × 15	3.6311	7.4092	7.4092	10.9276	13.3066	13.3596	17.0374	17.0374
		Yang and Shen (2002)	3.6636	7.2544	7.2544	10.3924	11.7054	12.3175	14.4520	-
		Diff. %	-0.8871	2.1339	2.1339	5.1499	14.1319	8.4603	17.8896	-
		Li et al. (2009a)	3.7202	7.3010	7.3010	10.3348	12.2256	12.3563	14.8112	14.8112
		Diff. %	-2.3950	1.4819	1.4819	5.7359	8.8421	8.1197	15.0305	15.0305
	10.0	8 × 8	3.6297	7.4159	7.4795	10.9530	13.8403	13.9777	17.0267	23.3436
		10 × 10	3.6297	7.4039	7.4039	10.9530	13.8403	13.9777	17.0261	17.0261
		13 × 13	3.6286	7.4039	7.4039	10.9367	13.3216	13.8813	17.0261	17.0261
		15 × 15	3.6285	7.4039	7.4039	10.9205	13.2969	13.3499	17.0261	17.0261
		Yang and Shen (2002)	3.1835	6.3001	6.3001	9.0171	10.2372	10.6781	12.6015	12.9948
		Li et al. (2009a)	3.1398	6.1857	6.1857	8.7653	10.3727	10.4866	12.5971	12.5971

Source: Reproduced from Chakraverty, S., Pradhan, K.K., Free vibration of exponential functionally graded rectangular plates in thermal environment with general boundary conditions. Aerospace Science and Technology 2014;36:132–156.

Table 8.20 Convergence of the first eight natural frequencies of CCCC exponential FG (SUS304/Si_3N_4) plates with different thermal conditions ($a = 0.2$ m, $h/b = 0.1$, $T = 300$ K, $\Delta T = 300$ K)

a/b (thermal state)	Source	λ_1	λ_2	λ_3	λ_4	λ_5	λ_6	λ_7	λ_8
0.5 (uniform rise)	8 × 8	9.9168	12.8597	18.4289	25.8804	29.0383	52.4360	55.0390	90.4605
	10 × 10	9.9168	12.8480	18.4289	25.8365	27.2488	29.0383	33.9022	52.4360
	13 × 13	9.9158	12.8480	18.4178	25.8365	27.2488	28.7179	33.9022	49.8631
	15 × 15	9.9153	12.8480	18.0903	25.8365	27.2488	28.7068	33.9022	40.0276
0.5 (Linear rise)	8 × 8	9.9296	12.8750	18.4499	25.9118	29.0712	52.4979	55.1015	90.5662
	10 × 10	9.9296	12.8633	18.4499	25.8679	27.2794	29.0712	33.9382	52.4979
	13 × 13	9.9286	12.8633	18.4388	25.8679	27.2794	28.7505	33.9382	49.9223
	15 × 15	9.9282	12.8633	18.1108	25.8679	27.2794	28.7393	33.9382	40.0727
1.0 (uniform rise)	8 × 8	3.6337	7.4241	7.4878	10.9642	13.8561	13.9936	17.0442	23.3707
	10 × 10	3.6337	7.4121	7.4121	10.9642	13.8561	13.9936	17.0437	17.0437
	13 × 13	3.6326	7.4121	7.4121	10.9477	13.3367	13.8972	17.0437	17.0437
	15 × 15	3.6326	7.4121	7.4121	10.9316	13.3120	13.3651	17.0437	17.0437
1.0 (Linear rise)	8 × 8	3.6376	7.4319	7.4958	10.9749	13.8714	14.0088	17.0612	23.3968
	10 × 10	3.6376	7.4199	7.4199	10.9749	13.8714	14.0088	17.0607	17.0607
	13 × 13	3.6365	7.4199	7.4199	10.9584	13.3513	13.9125	17.0607	17.0607
	15 × 15	3.6364	7.4199	7.4199	10.9422	13.3265	13.3796	17.0607	17.0607

Source: Reproduced from Chakraverty, S., Pradhan, K.K., Free vibration of exponential functionally graded rectangular plates in thermal environment with general boundary conditions. Aerospace Science and Technology 2014;36:132–156.

Table 8.21 First eight non-dimensional frequencies of CCCC exponential FG ($SUS304/Si_3N_4$) plate subjected to uniform temperature rise ($a = 0.2$ m, $T = 300$ K)

a/b	h/b	ΔT	λ_1	λ_2	λ_3	λ_4	λ_5	λ_6	λ_7	λ_8
0.5	0.1	0	9.7888	12.6982	17.9305	25.7085	27.1141	28.5726	33.7716	39.9238
		300	9.6126	12.4895	17.6933	25.4651	26.8669	28.3191	33.5096	39.6563
		500	9.4931	12.3483	17.5334	25.3015	26.7009	28.1489	33.3338	39.4769
	0.2	0	9.9043	12.8350	18.0822	25.8507	27.2603	28.7210	33.9221	40.0661
		300	9.8465	12.7664	17.9997	25.7515	27.1613	28.6179	33.8122	39.9423
		500	9.8078	12.7205	17.9444	25.6851	27.0951	28.5491	33.7387	39.8596
	0.25	0	9.9180	12.8514	18.1004	25.8677	27.2778	28.7388	33.9401	40.0832
		300	9.8742	12.7992	18.0361	25.7856	27.1964	28.6536	33.8483	39.9765
		500	9.8448	12.7643	17.9931	25.7308	27.1419	28.5967	33.7869	39.9053
1.0	0.1	0	3.4699	7.2307	7.2307	10.7436	13.1288	13.1854	16.8703	16.8703
		300	3.2831	7.0144	7.0144	10.5126	12.8889	12.9493	16.6329	16.6329
		500	3.1518	6.8662	6.8662	10.3556	12.7265	12.7895	16.4728	16.4728
	0.2	0	3.5981	7.3781	7.3781	10.8995	13.2863	13.3403	17.0236	17.0236
		300	3.5491	7.3153	7.3153	10.8289	13.2081	13.2628	16.9424	16.9424
		500	3.5161	7.2732	7.2732	10.7816	13.1557	13.2109	16.8881	16.8881
	0.25	0	3.6132	7.3956	7.3956	10.9180	13.3051	13.3588	17.0419	17.0419
		300	3.5796	7.3505	7.3505	10.8662	13.2458	13.2999	16.9792	16.9792
		500	3.5570	7.3204	7.3204	10.8316	13.2062	13.2606	16.9373	16.9373

Source: Reproduced from Chakraverty, S., Pradhan, K.K., Free vibration of exponential functionally graded rectangular plates in thermal environment with general boundary conditions. Aerospace Science and Technology 2014;36:132–156.

Table 8.22 First eight non-dimensional frequencies of CCCC exponential FG ($SUS304/Si_3N_4$) plate subjected to linear temperature rise ($a = 0.2$ m, $T = 300$ K)

a/b	h/b	ΔT	λ_1	λ_2	λ_3	λ_4	λ_5	λ_6	λ_7	λ_8
0.5	0.1	0	9.7888	12.6982	17.9305	25.7085	27.1141	28.5726	33.7716	39.9238
		300	9.7073	12.6016	17.8208	25.5962	27.0000	28.4557	33.6508	39.8008
		500	9.6525	12.5368	17.7472	25.5211	26.9237	28.3775	33.5701	39.7186
	0.2	0	9.9043	12.8350	18.0822	25.8507	27.2603	28.7210	33.9221	40.0661
		300	9.8778	12.8036	18.0445	25.8056	27.2153	28.6742	33.8722	40.0102
		500	9.8601	12.7825	18.0192	25.7756	27.1853	28.6429	33.8389	39.9729
	0.25	0	9.9180	12.8514	18.1004	25.8677	27.2778	28.7388	33.9401	40.0832
		300	9.8980	12.8276	18.0711	25.8306	27.2411	28.7003	33.8987	40.0353
		500	9.8847	12.8117	18.0516	25.8059	27.2165	28.6747	33.8711	40.0033
1.0	0.1	0	3.4699	7.2307	7.2307	10.7436	13.1288	13.1854	16.8703	16.8703
		300	3.3841	7.1307	7.1307	10.6367	13.0178	13.0761	16.7604	16.7604
		500	3.3255	7.0633	7.0633	10.5648	12.9433	13.0028	16.6868	16.6868
	0.2	0	3.5981	7.3781	7.3781	10.8995	13.2863	13.3403	17.0236	17.0236
		300	3.5755	7.3492	7.3492	10.8670	13.2504	13.3048	16.9865	16.9865
		500	3.5603	7.3298	7.3298	10.8453	13.2265	13.2810	16.9616	16.9616
	0.25	0	3.6132	7.3956	7.3956	10.9180	13.3051	13.3588	17.0419	17.0419
		300	3.5977	7.3749	7.3749	10.8943	13.2781	13.3319	17.0134	17.0134
		500	3.5873	7.3611	7.3611	10.8785	13.2600	13.3140	16.9943	16.9943

Source: Reproduced from Chakraverty, S., Pradhan, K.K., Free vibration of exponential functionally graded rectangular plates in thermal environment with general boundary conditions. Aerospace Science and Technology 2014;36:132–156.

Table 8.23 First eight non-dimensional frequencies of SSSS exponential FG (SUS304/Si_3N_4) plate subjected to uniform temperature rise ($a = 0.2$ m, $T = 300$ K)

a/b	h/b	ΔT	λ_1	λ_2	λ_3	λ_4	λ_5	λ_6	λ_7	λ_8
0.5	0.1	0	4.7368	7.7528	12.8694	16.7878	19.8426	30.4575	30.8328	37.2910
		300	4.4581	7.4795	12.5934	16.5006	19.5552	30.1819	30.5772	36.9665
		500	4.2622	7.2916	12.4059	16.3064	19.3612	29.9969	30.4057	36.7485
	0.2	0	4.9289	7.9423	13.0567	16.9751	20.0293	30.6239	30.9792	37.4761
		300	4.8569	7.8671	12.9730	16.8795	19.9323	30.5169	30.8718	37.3389
		500	4.8083	7.8166	12.9169	16.8154	19.8673	30.4455	30.8000	37.2472
	0.25	0	4.9515	7.9648	13.0789	16.9974	20.0516	30.6439	30.9967	37.4983
		300	4.9026	7.9124	13.0178	16.9244	19.9771	30.5569	30.9069	37.3834
		500	4.8697	7.8772	12.9769	16.8755	19.9272	30.4989	30.8470	37.3066
1.0	0.1	0	1.7312	4.7541	4.7541	7.7815	9.8815	9.8830	16.8528	16.8528
		300	1.4121	4.4761	4.4761	7.5088	9.6060	9.6075	16.6062	16.6062
		500	1.1513	4.2808	4.2808	7.3213	9.4178	9.4194	16.4398	16.4398
	0.2	0	1.9339	4.9458	4.9458	7.9706	10.0689	10.0704	17.0129	17.0129
		300	1.8655	4.8739	4.8739	7.8954	9.9876	9.9891	16.9297	16.9297
		500	1.8185	4.8253	4.8253	7.8449	9.9330	9.9345	16.8740	16.8740
	0.25	0	1.9568	4.9683	4.9683	7.9929	10.0912	10.0926	17.0321	17.0321
		300	1.9127	4.9194	4.9194	7.9406	10.0324	10.0339	16.9681	16.9681
		500	1.8827	4.8866	4.8866	7.9054	9.9931	9.9945	16.9254	16.9254

Source: Reproduced from Chakraverty, S., Pradhan, K.K., Free vibration of exponential functionally graded rectangular plates in thermal environment with general boundary conditions. Aerospace Science and Technology 2014;36:132–156.

Table 8.24 First eight non-dimensional frequencies of SSSS exponential FG (SUS304/Si_3N_4) plate subjected to linear temperature rise ($a = 0.2$ m, $T = 300$ K)

a/b	h/b	ΔT	λ_1	λ_2	λ_3	λ_4	λ_5	λ_6	λ_7	λ_8
0.5	0.1	0	4.7368	7.7528	12.8694	16.7878	19.8426	30.4575	30.8328	37.2910
		300	4.6087	7.6265	12.7416	16.6549	19.7096	30.3302	30.7149	37.1415
		500	4.5214	7.5412	12.6557	16.5658	19.6204	30.2451	30.6362	37.0415
	0.2	0	4.9289	7.9423	13.0567	16.9751	20.0293	30.6239	30.9792	37.4761
		300	4.8956	7.9076	13.0181	16.9313	19.9849	30.5753	30.9305	37.4139
		500	4.8732	7.8843	12.9924	16.9020	19.9552	30.5428	30.8979	37.3724
	0.25	0	4.9515	7.9648	13.0789	16.9974	20.0516	30.6439	30.9967	37.4983
		300	4.9289	7.9406	13.0509	16.9641	20.0177	30.6045	30.9563	37.4465
		500	4.9138	7.9245	13.0322	16.9419	19.9950	30.5783	30.9293	37.4119
1.0	0.1	0	1.7312	4.7541	4.7541	7.7815	9.8815	9.8830	16.8528	16.8528
		300	1.5900	4.6264	4.6264	7.6555	9.7541	9.7556	16.7387	16.7387
		500	1.4885	4.5392	4.5392	7.5703	9.6682	9.6697	16.6621	16.6621
	0.2	0	1.9339	4.9458	4.9458	7.9706	10.0689	10.0704	17.0129	17.0129
		300	1.9023	4.9125	4.9125	7.9358	10.0315	10.0329	16.9749	16.9749
		500	1.8809	4.8902	4.8902	7.9126	10.0064	10.0078	16.9494	16.9494
	0.25	0	1.9568	4.9683	4.9683	7.9929	10.0912	10.0926	17.0321	17.0321
		300	1.9364	4.9457	4.9457	7.9688	10.0642	10.0657	17.0029	17.0029
		500	1.9226	4.9306	4.9306	7.9527	10.0462	10.0476	16.9835	16.9835

Source: Reproduced from Chakraverty, S., Pradhan, K.K., Free vibration of exponential functionally graded rectangular plates in thermal environment with general boundary conditions. Aerospace Science and Technology 2014;36:132–156.

results evaluated in these tables, we may summarize the following related to thermal effects.

- Irrespective of the edge support and thermal state considered, frequency parameters are increasing with an increase in thickness-to-breadth (h/b) ratio, and are decreasing with an increase in the temperature gradient (ΔT).
- For the exponential FG plate with a certain edge condition and $\Delta T = 0$, non-dimensional frequencies coincide for the respective aspect ratio and thickness-to-breadth ratio assumed. For non-zero temperature gradients, natural frequencies at each mode associated with linear temperature rise are comparatively higher than that of the uniform case.
- It can also be noticed that the second non-dimensional frequency is coinciding with the third for FG square plate irrespective of the thermal condition and h/b considered. But no coincidence can be observed for natural frequencies of rectangular FG plates. This proposition may be true because of the fact that either the clamped or simply supported boundary condition is assumed at all the edges of the exponential FG plate.

8.3 PIEZOELECTRICITY

In the recent era, FG materials along with energy harvesting (EH) techniques have received crucial attention in engineering applications and therefore in the academic world. The requirements of these characteristics can be found in various potential applications, viz. biomedical devices, aerospace, textiles, civil engineering, bionics engineering, energy, electronic engineering, and micro-electro-mechanical systems (MEMS), etc. Past research involves the materials with functional gradation and nowadays a hot-cake in a variety of fields. The FG materials are generally lightweight composites which can afford severe temperature fluctuations across a mere interface. On the other hand, a very common EH material is 'piezoelectric,' which exhibits uncommon behavior. In 1880, Jacques and Pierre Curie discovered an unusual characteristic of certain crystalline minerals; when subjected to a mechanical force, these crystals became electrically polarized. Subsequently, the converse of this relationship was confirmed; if one of these voltage-generating crystals was exposed to an electric field, it lengthened or shortened according to the polarity of the field.

One of the basic vibration-based harvesting materials, the piezoelectric effect is the ability of certain materials to generate an electric charge in response to applied mechanical stress. One of the unique characteristics of the piezoelectric effect is that it is reversible, meaning that materials exhibiting the direct piezoelectric effect (the generation of electricity when stress is applied) also exhibit the converse piezoelectric effect (the generation of stress when an electric field is applied). When piezoelectric material is placed under mechanical stress, a shifting of the positive and negative charge centers in the material takes place, which then results in an external electrical field. However, an outer electrical field either stretches or compresses the piezoelectric material when the process is reversed. Piezoelectric materials are useful for vibration control as both an actuator and a sensor, and these have received much attention in various engineering applications. In structural applications, these actuators can be surface bonded or embedded. A short review of investigations performed previously are outlined below, along with governing equations associated with these problems.

Based on the theory of elasticity and piezoelectricity, the state-space-based differential quadrature method (SSDQM) is implemented to study free vibration of a functionally graded piezoelectric material (FGPM) under different BCs by Yang and Zhifei (2009). Bian et al. (2006) have found an exact analysis based on state-space formulation to study FG beams integrated with surface piezoelectric actuators and sensors. The plane stress problem of generally anisotropic piezoelectric beams is solved by Huang et al. (2007) with coefficients of elastic compliance, and piezoelectric and dielectric impermeability. Moreover, free vibration of statically thermal postbuckled FG material beams with surface-bonded piezoelectric layers subject to both temperature rise and voltage is investigated (Li et al., 2009b). Analytical solution for FG material beams integrated with piezoelectric actuator and sensor has been studied by Alibeigloo (2010) under an applied electric field and thermomechanical load. A size-dependent FG piezoelectric beam model is developed in Li et al. (2014) using a variational formulation to find the static bending and free vibration characteristics. On the other hand, one may see that a very few works have been done in the case of piezoelectric FG plates with different geometries. In this regard, Chen and Ding (2002) have derived three-dimensional theory equations of transversely isotropic piezoelectricity and two independent state equations with variable coefficients for FG piezoelectric rectangular plates. The nonlinear dynamic response and active vibration control of piezoelectric

FG plates are analyzed in Yiqi and Yiming (2010). Harmonic forced vibration of circular FG plates integrated with two uniformly distributed piezoelectric actuator faces has been studied by Jandaghian et al. (2014).

8.3.1 Equilibrium Equations

From the above literature survey, it is found that a very few investigations have been performed on vibration characteristics of FG beams and plates integrated with piezoelectric layers. As such, this section involves basic equilibrium equations related to these FG structural members.

FG Piezoelectric Beam

As given in Li et al. (2009b), let us consider an FGPM beam with length L and depth h, the beam is polarized in z direction and has a graded profile in this direction. $(\varepsilon_x, \varepsilon_z, \varepsilon_{xz})$, $(\sigma_x, \sigma_z, \tau_{xz})$, (D_x, D_z), and (E_x, E_z) denote the components of strain, stress, induction, and electric field, respectively. The two-dimensional constitutive equations for piezoelectric materials can be written as

$$\begin{Bmatrix} \sigma_x \\ \sigma_z \\ \tau_{xz} \end{Bmatrix} = \begin{pmatrix} C_{11} & C_{13} & 0 \\ C_{13} & C_{33} & 0 \\ 0 & 0 & C_{55} \end{pmatrix} \begin{Bmatrix} \varepsilon_x \\ \varepsilon_z \\ \varepsilon_{xz} \end{Bmatrix} - \begin{pmatrix} 0 & Q_{31} \\ 0 & Q_{33} \\ Q_{15} & 0 \end{pmatrix} \begin{Bmatrix} E_x \\ E_z \end{Bmatrix} \tag{8.16}$$

$$\begin{Bmatrix} D_x \\ D_z \end{Bmatrix} = \begin{pmatrix} 0 & 0 & Q_{15} \\ Q_{31} & Q_{33} & 0 \end{pmatrix} \begin{Bmatrix} \varepsilon_x \\ \varepsilon_z \\ \varepsilon_{xz} \end{Bmatrix} + \begin{pmatrix} G_{11} & 0 \\ 0 & G_{33} \end{pmatrix} \begin{Bmatrix} E_x \\ E_z \end{Bmatrix} \tag{8.17}$$

where

$$C_{11} = c_{11} - \frac{c_{12}c_{12}}{c_{22}}, \quad C_{13} = c_{13} - \frac{c_{23}c_{12}}{c_{22}}, \quad C_{33} = c_{33} - \frac{c_{13}c_{13}}{c_{22}},$$

$$Q_{31} = e_{31} - \frac{c_{12}e_{32}}{c_{22}}, \quad Q_{33} = e_{33} - \frac{c_{23}e_{32}}{c_{22}}, \quad Q_{15} = e_{15},$$

$$G_{11} = \beta_{11}, \quad G_{33} = \beta_{33} + \frac{e_{32}^2}{c_{22}}$$

in which c_{ij} or C_{ij}, e_{ij} or Q_{ij} as well as β_{ij} or G_{ij} are the components of the effective elastic, and piezoelectric and dielectric constants of the piezoelectric material in the three-dimensional or plane stress case, respectively. In addition, the components of strain and electric field can be expressed by using the displacement components $(u, \ w)$ and the electric potential (ϕ) as

$$\varepsilon_x = \frac{\partial u}{\partial x}, \quad \varepsilon_z = \frac{\partial w}{\partial z}, \quad \varepsilon_{xz} = \frac{\partial u}{\partial z} + \frac{\partial w}{\partial x}$$
$$E_x = -\frac{\partial \phi}{\partial x}, \quad E_z = -\frac{\partial \phi}{\partial z}$$

Without the consideration of body force and body charge, the dynamic equilibrium equations can be expressed as

$$\begin{aligned} \frac{\partial \sigma_x}{\partial x} + \frac{\partial \tau_{xz}}{\partial z} &= \rho \frac{\partial^2 u}{\partial t^2} \\ \frac{\partial \tau_{xz}}{\partial x} + \frac{\partial \sigma_z}{\partial z} &= \rho \frac{\partial^2 w}{\partial t^2} \\ \frac{\partial D_x}{\partial x} + \frac{\partial D_z}{\partial z} &= 0 \end{aligned} \tag{8.18}$$

where ρ and t denote the mass density and time, respectively.

FG Piezoelectric Plate

As explained by Chen and Ding (2002), the constitutive relations of a transversely isotropic piezoelectric medium in cartesian coordinates (with the z-axis being normal to the plane of isotropy) may be written as

$$\begin{aligned} \sigma_x &= c_{11}\frac{\partial u}{\partial x} + c_{12}\frac{\partial v}{\partial y} + c_{13}\frac{\partial w}{\partial z} + e_{31}\frac{\partial \phi}{\partial z} \\ \tau_{xz} &= c_{44}\left(\frac{\partial u}{\partial z} + \frac{\partial w}{\partial x}\right) + e_{15}\frac{\partial \phi}{\partial x} \\ \sigma_y &= c_{12}\frac{\partial u}{\partial x} + c_{11}\frac{\partial v}{\partial y} + c_{13}\frac{\partial w}{\partial z} + e_{31}\frac{\partial \phi}{\partial z} \\ \tau_{yz} &= c_{44}\left(\frac{\partial v}{\partial z} + \frac{\partial w}{\partial y}\right) + e_{15}\frac{\partial \phi}{\partial y} \\ \sigma_z &= c_{13}\frac{\partial u}{\partial x} + c_{13}\frac{\partial v}{\partial y} + c_{33}\frac{\partial w}{\partial z} + e_{33}\frac{\partial \phi}{\partial z} \\ \tau_{xy} &= c_{66}\left(\frac{\partial u}{\partial y} + \frac{\partial v}{\partial x}\right) \\ D_x &= e_{15}\left(\frac{\partial u}{\partial z} + \frac{\partial w}{\partial x}\right) - \varepsilon_{11}\frac{\partial \phi}{\partial x} \\ D_y &= e_{15}\left(\frac{\partial v}{\partial z} + \frac{\partial w}{\partial y}\right) - \varepsilon_{11}\frac{\partial \phi}{\partial y} \\ D_z &= e_{31}\frac{\partial u}{\partial x} + e_{31}\frac{\partial v}{\partial y} + e_{33}\frac{\partial w}{\partial z} - \varepsilon_{33}\frac{\partial \phi}{\partial z} \end{aligned} \tag{8.19}$$

where ϕ and D_i are the electric potential and components of the electric displacement; σ_i and τ_{ij} are the normal and shear stresses; u, v, and w are components of the mechanical displacement in the x-, y-, and z-directions; and c_{ij}, ε_{ij}, and e_{ij} are the elastic, dielectric, and piezoelectric constants, respectively. A relation $c_{66} = (c_{11} - c_{12})/2$ holds for transverse isotropy. In the absence of body forces and free charge density, the governing equations take the following form

$$\begin{aligned}
&\frac{\partial \sigma_x}{\partial x} + \frac{\partial \tau_{xy}}{\partial y} + \frac{\partial \tau_{xz}}{\partial z} = \varrho \frac{\partial^2 u}{\partial t^2} \\
&\frac{\partial \tau_{xy}}{\partial x} + \frac{\partial \sigma_y}{\partial y} + \frac{\partial \tau_{yz}}{\partial z} = \varrho \frac{\partial^2 v}{\partial t^2} \\
&\frac{\partial \tau_{xz}}{\partial x} + \frac{\partial \tau_{yz}}{\partial y} + \frac{\partial \sigma_z}{\partial z} = \varrho\rho \frac{\partial^2 w}{\partial t^2} \\
&\frac{\partial D_x}{\partial x} + \frac{\partial D_y}{\partial y} + \frac{\partial D_z}{\partial z} = 0
\end{aligned} \tag{8.20}$$

Here, the material density ϱ is a function of z. The equilibrium equations for FG piezoelectric plate with other regular geometries along with complicating environments are very scarce in open resources, so these are not included here.

8.3.2 Significance of Piezoelectricity

In particular, the piezoelectric effect is very useful in variety of applications that involve the production and detection of sound, generation of high voltages, electronic frequency generation, microbalances, and ultra-fine focusing of optical assemblies. It is also the basis of a number of scientific instrumental techniques and everyday uses such as acting as the ignition source for cigarette lighters and push-start propane barbecues (Fig. 8.8).

In real life, the piezo-FG structural members play crucial roles in engineering design and structural applications. The following are a few of the beneficial aspects of piezoelectricity:

- It is interesting to note that wastage of energy is a big issue in recent years. The recycling of energies to different constructive forms using EH techniques will be an important solution in this era. For example, piezoelectric patches are commonly used EH devices which are being used in pedestrian streets of Japan to remove the direct utilities of electricity.

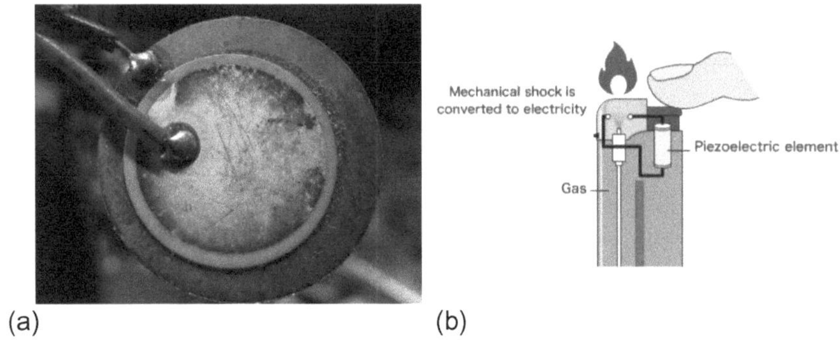

Figure 8.8 Utilities of piezoelectricity in daily life applications: (a) Piezoelectric buzzer; (b) Cigarette lighter. (*From Google Images*).

- Functionally graded materials are generally thermal resistant materials and, with the combination of EH materials, would be very useful in engineering design, architecture, and also in biological systems.
- The use of EH devices can be observed in several MEMS, which is a technology that combines computers with tiny mechanical devices such as sensors, valves, gears, mirrors, and actuators embedded in semiconductor chips.
- Most importantly, the scope of structural analysis and design is also to study their instabilities and failures due external stimulus, loads, and environmental hazards. It is worth eradicating these structural instabilities in engineering design by analyzing using various computational techniques.

8.4 CONCLUDING REMARKS

The present investigation has dealt with free vibration of FG plates under various complicating environments. The vibration of FG rectangular plates with power-law variation based material properties is to be handled under Winkler and Pasternak elastic foundations, whereas thermal environment is assumed to study the vibration characteristics of exponential FG plates. The generalized eigenvalue problem in each case can be found by using the Rayleigh-Ritz method, which is one of the most computationally efficient procedures to handle such problems. Effect of elastic foundation moduli and temperature gradients on free vibration characteristics of the plates may be summarized as follows.

- Considering the effect of the Winkler elastic foundation, it may be observed that the non-dimensional frequencies are increasing with an increase in aspect ratios for a fixed power-law index and are decreasing with an increase in power-law exponents for a fixed aspect ratio. This behavior is in fact true irrespective of the edge support, gradation criteria, and configuration of the FG plate (or in the presence of the Winkler elastic foundation) considered.
- A very interesting result may be seen for the effect of ascending values of power-law index (k) on free vibration characteristics of FG plates, which mainly depends on the presence of elastic foundations. Without elastic foundation, the eigenfrequencies follow descending behavior. They are also decreasing while considering the Winkler elastic foundation except for SFSF FG plates, where fundamental frequencies follow fluctuating behavior with an increase in k. Moreover, in the presence of either Pasternak or Winkler-Pasternak foundations, the frequencies may also decrease for SCSC Lévy plates, whereas the behavior for the other five combinations of edge conditions may not be straightforward.
- On the other hand, in the presence of Winkler and Pasternak foundations, the non-dimensional frequencies are gradually decreasing with an increase in values of E_r and are increasing with ascending values of ρ_r, which can be noticed from the expression concerned of λ.
- While looking into the results obtained for the exponential FG plate with thermal environments, it is worth mentioning that frequency parameters are increasing with an increase in thickness-to-breadth (h/b) ratio and are decreasing with an increase in the temperature gradient (ΔT), regardless of the edge condition and thermal state considered.
- Consequently, other complicating effects can also be taken into consideration to check the vibration characteristics of FG plates with different geometries.

CHAPTER 9

Practical Examples and Experimental Studies

Following than the theoretical and computational approaches given in previous chapters, it is also important to understand the experimental aspects of the related problems. This chapter deals especially with the practical implementation and experimental studies concerned with vibration characteristics of a few of the investigations done in the case of isotropic and functionally graded (FG) structural members. Different literature positions related to such problems certainly describe the utilities and experimental investigations of functionally graded materials (FGMs).

9.1 PRACTICAL APPLICATIONS

In this section, the real-life application of FGMs in a wide variety of fields have been outlined with the help of previous literature as below.

- Jedamzik et al. (2000): Given a novel processing route that is capable of manufacturing composites with continuous gradients in chemical composition based on electrochemical gradation and further infiltration of porous preforms.
- Bogdanski et al. (2002): Investigated the biocompatibility of nickel-titanium alloys by single-culture experiments on FG samples with a stepwise change in composition from pure nickel to pure titanium, including an Ti-Ni shape memory alloy for a 50:50 mixture.
- Abanto-Bueno and Lambros (2002): Investigated the crack growth resistance behavior of FGMs using full-field measurement technique of Digital Image Correlation (DIC). These FGMs are manufactured using ultraviolet irradiation of a photosensitive polyethylene copolymer.
- Liu and DuPont (2003): Fabricated crack-free FG TiC/Ti composite materials by laser engineered net shaping (LENS), with compositions changing from pure Ti to approximately 95 vol% TiC.

http://dx.doi.org/10.1016/B978-0-12-804228-1.00009-2

- Fu et al. (2004): Addressed a critical review on TiNi thin film based micro-actuators and its utilities in the rapidly growing field of micro-electro-mechanical systems (MEMS).
- Watari et al. (2009): Fabricated titanium/hydroxyapatite (Ti/HAP) and other FGM implants with concentration changed gradually in longitudinal direction of cylindrical shape by powder metallurgy to optimize both the mechanical properties and biocompatibilities in each region for biomedical implant application.
- Pompe et al. (2009): Provided an appropriate FG material, so-called hydroxyapatite (HAP)-collagen I scaffold, for in vitro growth of bone. This can be done in various routes, viz. sintering of laminated structures of HAP tapes filled with polymer spheres or combining biodegradable polysters with carbonated nanocrystalline hydroxyapatite.

9.2 EXPERIMENTAL STUDIES

The experimental investigation associated with a few models of structural members have been summarized here.

- Li et al. (2000) have described the fracture testing of model FGMs by using a hybrid numerical-experimental method. The FGMs are created by selective ultraviolet (UV) irradiation of a polymer, viz. ethylene carbon monoxide copolymer (ECO). The stress versus crack length relation is obtained experimentally and was used as boundary conditions at each increment of crack growth in a finite element analysis (FEA). Experimental setup for fracture resting of ECO and the sequence of CCD frames (a Sony XC-77 CCD camera and a Sanyo TLS-7000 VCR are used in parallel to record the crack initiation and growth events) describe the quasi-static growth in FGMs and are demonstrated clearly in this analysis.
- To overcome technical problems of structure control due to a large model size, Shi et al. (2004) have proposed an accurate finite element model; this is then verified by experiments on modal analysis. Firstly, numerical simulation of a real cantilever Euler-Bernoulli beam with active constrained layer damping (ACLD) is performed to analyze its model reduction process, stability, controllability, and observability. Then the modal experiments are carried out to verify this model and a real-time control system is designed based on the verified model as well. In real-time control experiment, the transverse displacement at the free tip of the cantilever beam is measured by the non-contact current

vortex sensor (CVS). A typical measure transverse displacement response at the free tip of the uncontrolled cantilever beam and corresponding controlled response have been estimated in this model.

- The theory governing the vibration of beams treated with constrained layer damping (CLD) that has FG viscoelastic cores is proposed by El-Sabbagh and Baz (2006) using the finite element method (FEM), and the predictions are verified experimentally for plain beams, beams treated with conventional CLD, and beams with CLD/FGVEM (FG visco elastic materials) of different configurations. Here, the experimental frequency response of the different tested configurations of the beam/CLD/FGVEM can be estimated, when subjected to a swept sinusoidal excitation over the frequency range between 0 and 400 Hz. The results estimated based on the theoretical predictions of the finite element model are validated with the experimental results, which seems to be in good agreement. These facts are neatly described in this literature.
- In common practice, concrete used in the field suffers from lack of durability and homogeneity. For the sake of economy, strength, and anti-corrosion, functionally graded beams (FGBs) have been developed with one layer of normal concrete and another of high volume fly-ash concrete (HVFAC). In this regard, Bajaj et al. (2014) have analyzed the flexural behavior of FGB with experiments and also with variation in interface as 0, 25, 50, 75, and 100 from bottom. In this study, HVFAC is prepared with replacement of cement by 20%, 35%, and 55% with fly ash for M20 and M30 grade of concrete, which yield very interesting properties, viz. 12.86% and 3.56% increase in compressive and flexural strength of FGB and optimized bond strength at 50 mm depth.
- Srinivasa et al. (2014) have performed experimental and finite element studies on free vibration of isotropic and laminated composite cylindrical skew panels. Natural frequencies were determined using CQUAD8 finite element of MSC Nastran and validated with experimental values. The experimental setup considers the test specimen with two opposite edges fully clamped and the remaining two edges completely free; the piezoelectric accelerometer was directly mounted on the test specimen at the geometric center using an adhesive. The accelerometer was then connected to signal-conditioning unit (fast Fourier transform analyzer), where the signal goes through the charge amplifier and an analog-to-digital converter, and the panel was excited at a selected point by means of impact hammer. This analysis has certainly shown the variation

of non-dimensional frequency coefficient (K_f) for isotropic cylindrical skew panels along with the variation of K_f with the aspect ratio and laminate stacking sequence for two different values of skew angles ($\alpha = 0°, 45°$). It can easily be said that the experimental values of the first, second, and third natural frequencies are in good agreement with the finite element solution in the case of both isotropic and laminated composite cylindrical skew panels.

In addition, it is worth observing that abbreviated notions in schematic experimental setups can be found in respective literature. In view of the above, one may find various literature positions for such practical uses and experimental validation of different models. It is worth mentioning that very few are available in the case of vibration of FG beams and plates. As such, researchers need to undertake experimental investigation related to FG structural members, which may be an open area of research. The numerical results provided in various chapters for different FG beams and plates may be used as benchmark results to validate the experimental results.

FSC
www.fsc.org
MIX
Papier | Fördert
gute Waldnutzung
FSC® C083411

Kolibri 360 GmbH
Ferdinand-Jühlke-Straße 7
99095 Erfurt, Deutschland
produktsicherheit@kolibri360.de